2/28/90

... *d*) ... Perspectives in Theoretical and

RANDLES and WOLFE • Introduction to the Theory of Nonparametric Statistics
RAO • Linear Statistical Inference and Its Applications, *Second Edition*
RAO • Real and Stochastic Analysis
RAO and SEDRANSK • W.G. Cochran's Impact on Statistics
RAO • Asymptotic Theory of Statistical Inference
ROHATGI • An Introduction to Probability Theory and Mathematical Statistics
ROHATGI • Statistical Inference
ROSS • Stochastic Processes
RUBINSTEIN • Simulation
SCHEFFE • The Analysis of
SEBER • Linear Regression
SEBER • Multivariate Observ
SEN • Sequential Nonparame ... Statistical Inference
SERFLING • Approximation ... of Mathematical Statistics
SHORACK and WELLNER • Empirical Processes with Applications to Statistics
TJUR • Probability Based on Radon Measures
ROGERS and WILLIAMS • Diffusions, Markov Processes, and Martingales, Volume II: Itô Calculus

Applied Probability and Statistics
ABRAHAM and LEDOLTER • Statistical Methods for Forecasting
AGRESTI • Analysis of Ordinal Categorical Data
AICKIN • Linear Statistical Analysis of Discrete Data
ANDERSON and LOYNES • The Teaching of Practical Statistics
ANDERSON, AUQUIER, HAUCK, OAKES, VANDAELE, and WEISBERG • Statistical Methods for Comparative Studies
ARTHANARI and DODGE • Mathematical Programming in Statistics
ASMUSSEN • Applied Probability and Queues
BAILEY • The Elements of Stochastic Processes with Applications to the Natural Sciences
BAILEY • Mathematics, Statistics and Systems for Health
BARNETT • Interpreting Multivariate Data
BARNETT and LEWIS • Outliers in Statistical Data, *Second Edition*
BARTHOLOMEW • Stochastic Models for Social Processes, *Third Edition*
BARTHOLOMEW and FORBES • Statistical Techniques for Manpower Planning
BECK and ARNOLD • Parameter Estimation in Engineering and Science
BELSLEY, KUH, and WELSCH • Regression Diagnostics: Identifying Influential Data and Sources of Collinearity
BHAT • Elements of Applied Stochastic Processes, *Second Edition*
BLOOMFIELD • Fourier Analysis of Time Series: An Introduction
BOX • R. A. Fisher, The Life of a Scientist
BOX and DRAPER • Empirical Model-Building and Response Surfaces
BOX and DRAPER • Evolutionary Operation: A Statistical Method for Process Improvement
BOX, HUNTER, and HUNTER • Statistics for Experimenters: An Introduction to Design, Data Analysis, and Model Building
BROWN and HOLLANDER • Statistics: A Biomedical Introduction
BUNKE and BUNKE • Statistical Inference in Linear Models, Volume I
CHAMBERS • Computational Methods for Data Analysis
CHATTERJEE and PRICE • Regression Analysis by Example
CHOW • Econometric Analysis by Control Methods
CLARKE and DISNEY • Probability and Random Processes: A First Course with Applications, *Second Edition*
COCHRAN • Sampling Techniques, *Third Edition*
COCHRAN and COX • Experimental Designs, *Second Edition*
CONOVER • Practical Nonparametric Statistics, *Second Edition*

(*continued on back*)

(continued from front)

The Teaching of Practical Statistics

The Teaching of Practical Statistics

C. W. ANDERSON

R. M. LOYNES

University of Sheffield

JOHN WILEY & SONS

Chichester · New York · Brisbane · Toronto · Singapore

Copyright © 1987 by John Wiley & Sons Ltd.

Library of Congress Cataloging-in-Publication Data:
Anderson, C. W.
   The teaching of practical statistics.
   (Wiley series in probability and mathematical
statistics)
   Bibliography: p.
   Includes indexes.
   1. Mathematical statistics—Study and teaching.
I. Loynes, R. M.   II. Title.   III. Series.
QA276.18.A53   1987   519.5   87–8238

ISBN 0 471 91572 6

British Library Cataloguing in Publication Data:
Anderson, C. W.
   The teaching of practical statistics.
   (Wiley series in probability and
mathematical statistics).
   1. Mathematical statistics—Study and teaching
   I. Title   II. Loynes, R. M.
   519.5'024372   QA276.18

ISBN 0 471 91572 6

Typeset by Macmillan India Ltd, Bangalore 25
Printed in Great Britian by Page Brothers Ltd, Norwich

Contents

Preface

Statistics is a subject with several aspects: there are various basic concepts (some of which remain controversial); there are methods based on these concepts; there are various arbitrary but convenient definitions, practices and so on; and, since in the end the subject is about understanding real or hypothetical data, there are skills and conventions concerned with relating the other aspects to data. At different times and in different circumstances fashion has led to changes in the emphasis that each of these has received within formal instruction, but the last – concern with the skills needed to apply theoretical knowledge to real problems – has never received much real attention, though it has had a certain amount of lip-service paid to it: to a great extent practitioners have had to learn the hard way. Quite apart from trivial considerations like fashion, there are of course a number of seemingly good reasons why practical skills should be made to take second place to other aspects: it is difficult to exercise practical skills without something to exercise them on; formal training schemes do not lend themselves easily to activities which need the subtle skills of practice; it is more important for most purposes to train the mind well – in logical reasoning and in the controlled use of imagination – than to take care over what may seem to be rather mundane concerns; it may be agreed that practical skills are essentially ones of maturity as compared with the more intellectual ones; and so on. But in fact at least some of these points either beg the question (is it beyond question more important to train the mind to the exclusion of other things?) or pose a false dichotomy (is it in fact necessary to make a choice between training the mind and attending to questions of practice – or can one get both? It probably *is* true that some technical *knowledge* has to be sacrificed if practical matters are to be dealt with, but that is not the same as saying that therefore the mind is receiving less training.)

Whether or not others are persuaded of this, however, we are quite convinced that no statistical training of other than the briefest and most casual kind ought to ignore the problem of making real connections with practice. This book is the outcome of this belief, which has involved us for many years in a search for ways to make such a connection and to a good deal of experimenting in our teaching at the University of Sheffield. Its contents may be thought of as being divided into three components. Chapters 1 and 2 discuss at some length what statistics is, and what are its special features, and end by putting forward a list of abilities that an ideal statistician would have. With this list as a point of reference, the various methods that have been used to promote the acquisition of practical skills are discussed in Chapter 3. Finally, we describe in Chapters 4 and 5 the method that we think comes nearest at present to meeting the requirements – a sequence of projects carefully tailored to the background and needs of the classes concerned.

Two points of a negative kind need to be made here. Firstly, although for definiteness the discussion is sometimes related to a particular school of inference (there are references to confidence intervals, for example), the essence of the book is quite unaffected by details of this kind. Secondly, we are not advocating an approach to the discovery of statistical concepts through practical work – a case can be made for this, but it needs an entirely different justification.

The book is addressed principally to all who teach statistics, and not merely to teachers of statisticians: that is, to school teachers as well as to university teachers, and to teachers of service courses in statistics as well as to those teaching specialist courses. Most of it, except for Chapter 3, should also be of interest to students, at least at higher levels: it will provide a vantage point from which to view their studies; it will point to various skills which are useful for statistical practice; and it will show at least some of the kinds of real problems which are accessible to statistical enquiry. These aspects are important for all students but particularly so for students working largely by themselves.

In the interests of avoiding clumsiness of style we have sometimes used the masculine words he, him, and his to refer to the generic student. Nothing is intended by this, and indeed about half our classes have always been female.

Our thanks go to those who read early drafts of parts of the book, including R. M. Allen, R. A. Bradley, D. R. Cox, D. J. Finney, D. W. Flaxen, K. T. de Graft Johnson, M. J. R. Healy, D. A. Preece, A. A. Rayner, E. J. Snell, S. Tulya Muhika, and to many others with whom we have discussed these questions, in particular to our colleagues in Sheffield; to Betty Gowland, Pauline Hollow and Elizabeth Womack, who transformed our manuscript into typescript; to generations of students, from whom we have learnt much about the teaching of statistics to Charlotte Farmer and her colleagues at J. Wiley for their efficiency and helpfulness in the preparation of this book for the press; and finally to our wives and families for their constant support and encouragement throughout the work on this project.

Acknowledgements

We are indebted to the following for permission to reproduce copyright material:

The Controller of Her Majesty's Stationery Office

The University of London

Writers of two letters to the Editor of *The Times*

Professor K. D. Duncan

Sheffield City Museums

The Editor of *New Scientist*

The Editor of *Ergonomics*

The Editor of *Le Monde*

The Editors of the *Journal of the Royal Statistical Society*, Series A.

The Editor of *Nature*

CHAPTER 1

Statistics and the education of statisticians

This book grows out of a certain view of the nature of statistics, and an attitude to the purpose of teaching and studying the subject which reflects this view. In this chapter we set out to describe this view and attitude, and to argue that statistics as a subject is inseparable from its application and practice, that any statistical education should take account of this connection, and that the benefits from trying directly to teach statistical practice can be substantial.

1.1 THE STUDY OF STATISTICS

We begin by asking, for statistics, some fundamental questions which, sooner or later, implicitly or explicitly, confront most teachers of any subject: Why do people study my subject? – and, perhaps more fundamentally, Why *should* they study it?

A large number of people of course do study statistics – indeed the subject must be counted as one of the major twentieth-century growth areas in higher education. When Karl Pearson was appointed to the Galton Chair in Applied Statistics at University College London in 1911 his new department was the first of its kind in the world; there were six other members of staff and only a handful of students (Pearson, 1971). From those early beginnings the study of statistics in institutions of higher education has proliferated: nowadays in the UK alone about 800 students graduate each year with a statistical first-degree or postgraduate qualification (Royal Statistical Society, 1986) and for the whole world the number must run into thousands. Even a rough calculation suggests that there must be several thousand teachers of statistics in universities and other institutions of higher education around the world. The total effort expended on both the teaching and the learning of the subject is therefore enormous. How can so great an investment be justified? What makes people choose to study statistics? Any answer to these questions, going beyond the purely idiosyncratic, is inevitably bound up with the nature of statistics and its place in the modern world: what, then, is this place?

The subject is curiously difficult to define. Tentative short definitions either leave something out or are all-embracing: 'Statistics is the study of uncertainty' we are told – but what then of non-probabilistic experience? 'Statistics is the collection of ideas and techniques, helpful for analysing data.' But so too, it could be said, is much of science: the general laws of physics certainly help explain vast amounts of observational data, and so in a sense help us to analyse them, but we would not wish to claim physics as a mere subset of statistics. Of course, if a clearer distinction between 'analysis' and 'explanation' is drawn, the possibility of this kind of confusion would be reduced, but that is merely to relocate the difficulty. Rather than pursue such philosophical questions let us turn to an alternative way of assessing and justifying the place of statistics in the world in general and in higher education in particular – by describing rather than defining. Inevitably for a subject that touches so many others there are several aspects.

One, relating to the second of the 'definitions' above, is what might be called the cultural justification of statistics: that statistical ideas underlie in a general sense our scientific outlook on the world, or, differently expressed,

> that Statistics in its broadest sense is the matrix of all experimental science – and is consequently a branch of scientific method, if not Scientific Method itself; and, hence, that it transcends the application of the scientific method in sundry fields of specialization. The scientist should know Statistics as he knows logic and formal language for communicating his ideas.

> (Kendall, 1968)

R. A. Fisher evidently had something of the same idea in mind when he said in his Presidential address to the Royal Statistical Society in 1953 that

> [statistical thinking in the broad sense] formed the silent background of the greatest scientific advances of the nineteenth century

through, for example, the kinetic theory of gases and statistical mechanics, and the Darwinian revolution in biology; and

> Statistical Science [is] the peculiar aspect of human progress which [gives] the twentieth century its special character.

These are rather grand claims, and many statisticians, we suspect, though they would not wish to dispute them, would prefer to justify their subject a shade less cosmically by pointing to various more detailed and mundane ways in which it is useful in the world – by describing, that is, in general terms (but more specifically than above) what useful things statisticians do.

Statistical activities, of course, are legion, and to describe them all in any detail would be impossible. However, extending slightly a classification put forward by Healy (1973), we suggest that they can roughly be divided into four broad kinds of work, and that this rough classification – together with recognition of the broader cultural role of statistics – helps to explain the demand for statistically knowledgeable graduates, and hence the efforts that are devoted to their education. We emphasize that the four categories we have in mind are only broad groupings, that they undoubtedly overlap to some extent, and that the activities of any individual statistician may well span more than one of them.

The first category, and the oldest historically, is what Healy calls the work of the 'intelligence statistician': the collection, ascertainment and presentation of facts, often those of interest to Government: economic indicators, trade indices, population figures, and such like. The data in this kind of work are typically the whole population, and formal probability notions do not enter: it is rather connections with economics and other wider aspects of social affairs that are likely to be most to the fore in what the statistician does. The origins of Political Arithmetic, and through it of modern statistics, in the work of Graunt, Petty, Halley, and others during the seventeenth century, lie largely in the domain of intelligence statistics. Nowadays not only do much government statistics and much so-called data analysis fall into this category, but also many of the other less weighty activities associated in the public mind with the term 'statistics' – the statistics of the sports commentator, for instance, or the election pundit. Less obviously, some aspects of the compilation and use of large computer databases – the medical records of all patients in a health authority region, say – are arguably part of intelligence statistics.

A second major type of statistical activity is work carried out in connection with 'scientific' research; for example the work of a statistics department at an agricultural research station – Fisher's work at Rothamstead, say – or of an academic consultant collaborating with colleagues in another discipline, or of a commercial market researcher. 'Scientific' is taken here in the broad sense of 'using scientific methods' and 'increasing the stock of knowledge' rather than as limiting the subject of investigation to the traditional experimental sciences. Academic consultants in particular might well meet problems from a much wider range of areas, from almost any discipline within their institution, in fact. The variety of activities in scientific statistics is enormous – from formulating research proposals, designing experiments and constructing models to devising algorithms and analysing data. A distinction (though not a hard-and-fast one) from intelligence statistics might be found in the emphasis on experimental or observational data rather than on whole populations. Natural tools tend to be probability theory, mathematical and statistical models, and statistical inference.

In contrast to the 'scientific statistician' or, as Healy calls him, the 'experimental statistician', in whose philosophy the open-ended question and the maximizing of information are paramount, there stands the statistician who works, perhaps in the management services team of a manufacturing company, a bank or a public utility, with more immediate and precisely formulated goals in view, all connected with the optimizing or improving of the operation of the organization. Though the demands of his work are no less rigorous than those made on the scientific statistician, and many of the tools he uses are the same, the emphasis in his activity lies more towards the reaching of a decision – to scrap this production batch, to site the factory in such-and-such a place, to implement this particular inventory-control system with that particular setting of parameters – rather than towards the pursuit of knowledge for its own sake: he may analyse experimental data as does his scientific colleague, but only up to a point sufficient for prespecified needs and not to extract the last drop of information. In this kind of statistical work more than in either of the earlier two it is likely that the 'statistician' will labour in disguise under some other title – management scientist, operations researcher, quality control engineer, reliability adviser, etc. Nevertheless we argue below that it is important to recognize him, at least for educational purposes, as just as legitimate a member of the species as any other kind of statistician.

The final type of statistical work is that of the theoretical statisticians, or, as Healy calls them, the statistical mathematicians. If we take the class to include probabilists then its work includes the refinement of mathematical techniques useful in statistics, theoretical examination of statistical methods, study of the foundations of statistical inference, development of new models in both abstract and applied probability, and the study of random systems generally. The demarcation line between 'theoretical' and 'scientific' statistics is particularly indistinct, but evidently the difference between the two is bound up to some extent at least with motivation: in 'scientific' statistics work is motivated fairly directly by a particular problem in some other substantive field of enquiry, whereas in 'theoretical' statistics motivation is either more remotely related to a single problem or comes from more than one problem – and perhaps a great many – in different areas. On this count Fisher's work on, say, the logarithmic series distribution might be classed as 'scientific' and his study of the maximum likelihood method as 'theoretical'. Both theoretical statistics and the non-theoretical kinds of statistical activity described earlier contribute crucially to the development of the subject as a whole, the one through the power that general approaches give and the other through the stimulus of new problems. W. Feller's (1968) recognition that the history of probability, and of mathematics more generally,

> . . . shows a stimulating interplay of theory and applications: theoretical progress opens new fields of applications, and in turn applications lead to new problems and fruitful research

applies with equal force to statistics too.

It may be objected that the categories of statistical activity above have been too broadly defined: that they include work more properly described as operations research, say, or econometrics, decision theory, data analysis, and so on. We know, too, of at least one distinguished probabilist who claims that he could never hope to aspire to the title of 'statistician'. How, then, can we justify casting the net so widely?

The reasons stem largely from educational considerations. All the kinds of work described above are in areas to which graduates with statistical training are recruited and in which they use their training. Most are linked with fields which were the historical cradle of the subject. Intelligence statistics, for example, whose place in modern statistics might be questioned on account of its largely non-probabilistic setting, and which is often neglected in current statistical teaching, is in a direct line of intellectual descent from the work of Graunt and others in the earliest days of the subject. The descent of many of the ideas – if not necessarily the context – of what might be called industrial statistics can similarly be traced back to pioneering work in insurance in the eighteenth century and earlier. At a fairly deep level there is a common thread running through the varied activities that we have labelled as statistical. What the thread is seems to come closest to being summed up in what Efron (1982) has called the fundamental idea of statistics:

> that useful information can be accrued from individual small bits of data.

On this count intelligence statistics is unquestionably part of statistics: and since mathematical probability is the theory that attempts to model that which is individually uncertain, probability theory and probability-based statistics are too. There is an intellectual coherence among the ideas that lie behind the activities we have listed, which makes it important at the educational level to treat these activities as aspects of the same subject. Much cross-fertilization of ideas and intellectual stimulus can be provided by an education that shows the development of a basic notion in a wide diversity of different contexts. (This is not to deny, of course, that different fields of applications also need *special* considerations and expertise, or that the student will not continue to learn about them on the job after the end of formal education.) At a more directly utilitarian level, too, such a catholic view of the subject fits well with the immediate vocational needs of most students, since most will not know until the end of their course which area of work they will find a job in. An education which conveys both the unity and the diversity of the subject gives an economical training for such a student and an overview of possibilities for future specialization.

The answer to the initial question about why effort should be devoted to the teaching and study of statistics is therefore clear. It is simply that the subject embodies ideas and methods which are extremely useful in a very wide variety of the problems facing mankind – from the organization of society and of industrial production and commercial activity to our attempts to understand the natural world. Society therefore needs people with a statistical training.

Two points of particular importance for statistical education emerge from this conclusion and the argument leading to it: firstly that statistics is inseparable from its application and practice, in the sense that it is ultimately justified by its usefulness in problems arising outside itself (and so is not fundamentally concerned with studying internal questions for their own sake); and secondly that the subject is broad, in contributing to and receiving stimulus from a very wide

range of other areas – wider perhaps than is recognized in some training programmes, where the subject as a whole has become identified with scientific and
theoretical statistics alone.

1.2 THE EDUCATION OF STATISTICIANS

If, as argued above, statistics is inseparable from its application and practice then
there are important implications for the education of statisticians.

Of course, like any academic discipline, statistics contains a systematized body
of concepts, theories and methods which have grown out of past experience and
which together form the intellectual framework within which its practitioners
operate. Much of this body of knowledge in statistics, as a numerical subject, is
inevitably mathematical in form. Any statistical education must introduce the
student to it and give him skill and self-confidence in its manipulation, but to
motivate the mathematics and to help keep in mind its place as a part of a larger
enterprise (the statistical analysis of real problems) it will surely be essential to
teach the theory with substantial regard to applications. How this is done will
vary with the particular piece of theory or methods being taught. The elementary
discussion of significance tests, for example, will almost always be accompanied
by some description of a typical problem in industry or scientific work whose
analysis could be furthered by thinking of it in terms of a test. A course on
'theoretical statistics' will typically be full of such small examples and would no
doubt appear rather mysterious without them; indeed, an ideal theoretical course
would make clear how all the ideas it studies grow out of practice. At the opposite
extreme whole courses may be devoted to one particular application or field of
application such as, for example, medical statistics or industrial statistics, and will
develop particular methods suited to the special needs of substantive problems
there. The whole course is then, in a sense, an illustration of the application of
statistical ideas. We might call what is taught in these kinds of course 'statistical
theory and method' and 'applied statistics' respectively (though the latter name
is often used for courses which are concerned only with statistical
methods – admittedly of use in the field of interest – rather than with applications
in the full sense of the word). Their contents form the backbone of the academic
subject of statistics, without which the status of the subject as a discipline in its
own right would be questionable.

Yet courses on these topics, however skilfully presented and fully understood,
do not supply the whole of the equipment that the statistician needs if he wishes to
progress beyond passive spectatorship of past achievements to a point where he
can use statistical ideas himself on real problems. Application of statistics usually
calls for inductive thought, and induction will inevitably have been de-
emphasized in the mathematically based, and necessarily deductive, setting of
courses on theory and methods. Indeed there is a curious tension here at the heart
of statistics, in that too whole-hearted an absorption of the ethos of the

mathematical language of the subject may actually stand in the way of its successful application. (This fact may be a partial explanation of the occasional regrettable inability of certain pure mathematicians and statisticians to understand and appreciate each other.) Mature statisticians will have reached an accommodation between these conflicting requirements of their craft, but the conflict poses peculiar difficulties for the student. In addition, problems to which statistical ideas are applied are usually complex – the richness of the real world rarely makes for tidiness. In consequence the fruitful application of statistical ideas is rarely automatic: as Tukey (1963) said:

> Who can believe that the fundamental process of learning about the world – the analysis of experience – can rest upon theories unflavoured by wisdom?.

The skilful statistician brings to bear on a problem not only his technical knowledge of statistical theory and methods, of the mathematics of probability, of properties of general mathematical/numerical techniques, and so on – and his awareness of how others have applied them in a variety of fields – but also a battery of skills and insights, working at an almost unconscious level, which guide his selection of appropriate technical tools and his use of them. By their nature these skills are difficult to define precisely. The practitioner uses them almost instinctively, and, like the top-class golfer or tennis-player asked to analyse his game, may well not be able to rationalize them at all: 'you just do it this way', he may say (and the more he thinks consciously about how he plays the stroke, the worse his performance becomes).

We can, however, attempt an indirect description of what is required, by considering the stages in the statistical analysis of some real problem. Of course it is not suggested that all statistical activity fits the following pattern, but only that this pattern provides a valid and useful parable. Suppose, then, that our statistician is operating as a consultant, and has hung out his sign and opened his door to anyone who wishes to approach him. Soon an outside investigator comes along with a problem. In the collaborative work which follows we might, adapting a description by Hunter (1981), distinguish the following stages:

(a) the investigator poses the problem and describes the background to it – in economics, medicine, administration, or whatever;
(b) the problem is formulated in statistical terms, possibly after considerable discussion and modification of its original form;
(c) the mathematical/statistical/computational machinery needed to produce a formal 'answer' is operated;
(d) the 'answer' is interpreted and evaluated in the light of the formal questions it was meant to answer;
(e) a response is communicated in the terms of the original question. ('Response' rather than 'answer', since if the original problem could not be answered with the data or resources available it would be important to say so.)

The distinction between (d) and (e) is meant to reflect the difference between technical internal checking – constancy of variance, say, or independence of errors in a regression analysis – and external substantive conclusions. The progression of a problem through stages (a) to (e) is rarely straightforward. As Hunter points out, the two fundamental processes of deciding how best to collect the necessary data for a particular study (assuming they are not already to hand) and how best to analyse them are exploratory, and require iteration from one to the other. Similarly, results and insights gained from an initial analysis may prompt a reformulation of the original question – a question about rate of change for example may need to be reformulated if the relationship turned out to be non-linear. For any real problem, therefore, there is inevitably much complicated looping, back-tracking and trial-and-error between stages on the way to a response.

The whole process demands much creative skill from the statistician: he will need:

● to synthesize ideas from his own statistical and mathematical background with those generated by the problem itself, and for this he will need a practical awareness of general scientific and wider issues as well as a thorough understanding of his own subject;
● to judge between alternative formulations and approaches;
● the technical facility to carry through whatever analysis he decides upon;
● good powers of interpretation and communication to evaluate and publicize the results.

Current practice in teaching emphasizes the knowledge-based formal aspects of these demands. Knowledge of statistical theory, methods and applications – which we have called the backbone of academic statistics – is, without question, central to the carrying out of almost any technical analysis that the statistician decides on, and it forms the indispensable basis for all the other skills he must exercise, but these other skills are not themselves part of knowledge; rather they are the techniques of putting knowledge to work for a specific purpose. Tukey calls them *wisdom*, echoing Francis Bacon, who in 1625 wrote of 'Studies', that they

> . . . teach not their own use; but that is a wisdom without them and above them, won by observation.

More prosaically we will refer to them in this book as the skills of statistical practice, or just 'practical statistics'. If the statistician is thought of as a craftsman whose purpose is to shape useful articles from the raw material of experience, then the principles, theories, and technical methods of statistics make up, in this analogy, the contents of his essential tool-kit, and 'practical statistics' consists of those abilities and tricks-of-the-trade which enable him to handle his tools skilfully and effectively, and which distinguish his skill from that of a novice in the same workshop.

An ideal education for a statistician would endow him with at least the rudiments of these skills to build on in his later career, together with the underlying technical knowledge on which they are based. The two-part nature of such an education would have a direct vocational benefit for the student, in easing his transition to statistical employment, and an educational benefit, in fostering a more complete understanding of statistics, including its theoretical aspects, which can come when the subject is seen in the context of its overall purpose of contributing to the solution of problems outside itself. Full appreciation of the utilitarian nature of the subject, and of the place of theory within it, would certainly increase the student's later scientific effectiveness.

An additional wider benefit, not specific to statistics, is that experience of the practical application of any abstract theory can help teach an approach to problems and encourage habits of thought which are of value to the student's general intellectual development, both within his studies and more widely. This is especially important in view of the evidence (Royal Statistical Society, 1986) that substantial numbers of statistical graduates find employment in fields remote from the subject.

1.3 THE TEACHING OF STATISTICS

How far do current programmes for the teaching of statistical specialists go towards providing the ideal education outlined in the previous section? We consider this question here for Bachelor's or Master's degree programmes in which the student is studying statistics as his main subject, with a view to a career as a statistician (in the broad sense). In Section 1.5 we discuss separately how far the ideal and attempts to attain it might be modified for those who are not specializing in statistics but who nevertheless study it to a greater or lesser extent along with other subjects.

The teaching of statistical principles and methods has for many years been an important part of training programmes for statisticians, and there is now broad agreement on what should be taught and on how it should be presented. Most programmes for example contain courses on:

- Probability theory – including some applied probability or stochastic processes.
- The mathematical theory of statistics – principles of inference, theories of estimation and hypothesis testing.
- Statistical methods – specific inference procedures, linear methods.

They often contain further topics such as:

sampling theory
design of experiments
time series analysis
analysis of categorical data
non-parametric and distribution-free methods
sequential methods

Most of these topics are covered by excellent textbooks which give the teacher ample foundations for his courses.

The teaching of the practical application of statistical ideas on the other hand is in a less satisfactory state. Although most substantial training programmes for statisticians include one or more 'applied statistics' courses or a 'practical class', there is nevertheless a fairly widespread feeling, especially among first employers of newly qualified students, that many programmes do not teach students how to practise statistics effectively. Evidence here is inevitably largely anecdotal, but two published reports which indirectly reflect the concern of employers are that of the American Statistical Association (ASA) Committee on the Training of Statisticians for Industry (1980), and, in the UK, that of a Joint Working Party of the Royal Statistical Society, the Institute of Statisticians and various groupings of academic statisticians, into the Supply of and Demand for Statisticians (Royal Statistical Society, 1986). The ASA Committee's report presents guidelines for programmes which train statisticians for industry, and strongly emphasizes the need to develop students' practical abilities: an implication is that not all programmes have yet succeeded in giving adequate emphasis to this aspect. The Joint Working Party found evidence that although there is apparently no shortage of statistical graduates in the UK in relation to the number of openings for them, nevertheless some employers find it difficult to recruit suitable candidates. The reasons for this phenomenon are unknown, but most of the likely explanations scarcely flatter the teaching of the subject; a lack of realism on the part of some employers may contribute to the perceptions of shortcomings – after all most other professionals, such as engineers, architects, and doctors, undergo some form of 'graduate apprenticeship' after finishing training and before being thought fully competent – but this does not seem enough to explain all the criticisms.

If it is true that the teaching of practical statistics is in a less satisfactory state than that of theory and methods then we do not have to look far for possible explanations. One important factor is that what needs to be taught under 'practical statistics' is much less easily definable, and so course objectives are not always clearly recognized. 'Practical' is easily confused with merely 'non-theoretical' or with 'applied' (in the sense of statistics in particular areas of application such as 'medical statistics' or 'agricultural statistics'). In the first connection any aspect of statistics which does not make direct use of algebra or analysis – anything numerical say, or to do with computers – may be taken to be 'practical'. Work nominally in 'practical statistics' can then cover an enormous range of activities – instruction in computer programming or in the use of packages, numerical illustrations of theoretical concepts (such as demonstrations of the central limit theorem by simulation, or numerical comparisons of the power of rival tests), 'experiments' with cards or dice or measuring instruments, routine numerical exercises – all of which are valuable in themselves, but largely for the help they give towards the consolidation of knowledge and to the understanding of theory and methods rather than for their immediate relevance to statistical practice. Similarly, the distinction between 'practical' and 'applied' has not always been clearly drawn. To learn about how statistics has been used in other fields is a vital part of the student's statistical education, but it is unlikely on its own fully to equip him for practice. Further, whatever it is that needs to be taught under the heading of 'practical statistics', it does not fit easily into the didactic pattern of much of classroom teaching. There is no doubt that it is easier to learn material that fits into a systematic formal structure, as statistical theory, method, and applications do, but practical statistics does not. Students and teachers of statistics necessarily move for much of their time in the thought-world of the logical structures of the mathematical theory, and that makes it understandable if other aspects of the subject, which depend on less consciously perceived mental processes, should sometimes remain relatively neglected. Indeed, in a discussion of the nature of scientific method, Box (1984) suggests that a division of mental activity such as that between the formal aspects of statistics on the one hand and practical statistics on the other, and a tendency for the former to dominate the latter, may be an inevitable reflection of features in the physiology of the human brain. Whatever the reason, these difficulties certainly exist, and they no doubt help to explain the undeniable fact that aids for the teaching of practical statistics, in the form of books and tested methods, are not so numerous as for the rest of the subject.

Part of the aim of the present book is to help fill this gap. In Chapter 2 we discuss in greater detail the qualities which one might hope would be developed by a form of teaching which aims explicitly at the practical application of Statistics. Chapter 3 reviews some of the methods that have been used for the teaching of statistical practice, and discusses their various strengths and weaknesses. One way of organizing such teaching is by means of a sequence of short student assignments. This provides a simple and effective framework within which the teacher can build on the strengths of particular teaching methods and the student can develop a wide range of practical abilities. In our view it represents the currently most effective way of teaching statistical practice. Chapter 4 is devoted to a fuller discussion of it, with examples; and, finally, Chapter 5 contains data and questions providing material for teaching within this framework.

1.4 FEASIBILITY AND DESIRABILITY

In Section 1.2 above we have suggested that some highly desirable benefits would accompany a statistical education which is successful in teaching the skills of statistical practice, but in Section 1.3 we have seen evidence that not all current programmes are fully effective in conveying these skills. The questions naturally arise as to whether the skills can actually be taught effectively, and, even if they can be taught, whether the effort required would be worth while. In this section we briefly discuss these questions and give reasons for our belief that the answer to both is 'yes'.

Arguments against attempting to teach the skills of practical statistics are that:

(a) They are not teachable: the student either possesses them, in the same way that he might possess, say, a sense of humour, or he is incapable of ever acquiring them. In either case the teacher is absolved of responsibility.

(b) They are best learnt through working experience in the first few years after the completion of formal training. The unpreparedness for practical work of graduates from statistical degree courses which nominally include practical training is sometimes cited as evidence for this view.

(c) There are so many demands on the limited time available in statistical programmes that it is more efficient to fill it with the maximum amount of abstract and theoretical material and to leave the relevance and practical application of this material to be worked out at leisure by the student during his subsequent career.

These views were expressed at a meeting of the Committee of Professors of Statistics in the UK and Ireland in 1977 (reported in Armitage, 1977) which discussed some of the same concerns about the effectiveness of current teaching as those referred to in Section 1.3.

Counter-arguments and reasons for believing that the skills of practical statistics are capable of being taught are that:

(i) Criticisms of the effectiveness of some existing statistical programmes, nominally containing practical training, should not be taken as conclusive evidence that practical skills can be learnt only on the job. As suggested in the previous section, some current practical training is poorly targetted and confuses practical statistics with applied statistics or with work having other purposes.

(ii) An analysis of the skills of practical statistics, detailed fully in Chapter 2, shows that many may be broken down into simpler abilities (those of effective communication and of working with others, for example) which are being successfully taught in other contexts, and which common sense suggests could be taught in, or in connection with, statistics too.

(iii) Most crucially, methods for teaching practical statistics have been developed, are in current use, and appear to be effective. Part of the aim of later chapters in this book is to present a detailed description and discussion of them. Their existence and effectiveness are a powerful argument against the first two objections to an attempt to teach practical statistics: it *can* be done.

The question of whether it should be done, however, remains. There is a price to pay for an expansion of the scope of a statistical education and for the introduction of new teaching methods. Part of this price lies in the intellectual effort that a reorientation of teaching requires of the teacher. Another part lies in the demands that the new activities make on the time of both the student and the teacher. None of the likely benefits from the new approaches can be expected to accrue without a substantial investment of effort, and it may well be that, as a result, less time will be available for the more traditional components of a statistical programme. The quality of an education cannot, however, be crudely measured in terms of the numbers of hours of exposure to different topics, and any nominal reduction in formal coverage of traditional areas is less serious than at first sight it might seem. The reoriented teaching can be expected to give the student both enhanced study-skills and also a deeper understanding of fundamental knowledge-based parts of the subject and their relation to practice, and these advantages will better equip him to make good any immediate gaps in his technical knowledge and, in the longer term, to keep abreast of the technical changes that are inevitable throughout his career. The modification in teaching is likely therefore to lead to a more efficient education. If on the other hand we do not modify statistical programmes so that they equip students more effectively to tackle substantive problems then there will also be a price to pay – and perhaps a larger one, at that. There is a substantial demand from industry, commerce, government, and scientific research for numerate personnel who can apply quantitative methods imaginatively to their problems. Many of these problems are statistical – in the broad sense discussed in Section 1.1 – or have statistical aspects, and so statistically trained people could be expected to meet a sizeable part of this demand. However, if statisticians find it difficult to put their knowledge into practice or to communicate their findings then potential employers will inevitably turn elsewhere for the expertise they need, to the detriment of the solution of many problems and of employment opportunities for statisticians. The result, feeding back into recruitment to teaching programmes, could be the gradual degeneration of the subject as a whole and its decline into a narrow academic backwater.

These reasons seem to us to make an overwhelming case for the vigorous teaching of statistical practice.

1.5 OTHER STUDENTS

The discussion up to now has been about the education of students specializing in statistics at first-degree or Master's degree level. We have suggested that an ideal education for these students would combine instruction in the principles, theories, and methods of statistics together with training directly aimed at developing the skills of practical statistics. Many students, however, although they study statistics to a greater or lesser extent during their education, do not specialize in it. We now ask how far the ideal and attempts to attain it might have to be modified to meet their needs.

Students in many other disciplines take elementary courses in statistics. Many continue their studies beyond the elementary level, and for some of them statistics may be a major subject at an advanced level, perhaps as part of a dual degree with another subject – mathematics, computing, economics or business studies, for instance. Students' reasons for studying statistics are similarly diverse. For some the aim is simply to obtain a broad appreciation of the subject's general concepts and scope. An introductory statistics course for medical students, for example, might aim to convey an appreciation of significance tests and p-values sufficient to enable the student to read, unperturbed, papers in the medical literature, but without equipping him with the knowledge to choose and carry out more than the most routine calculations himself. In such a course details of a particular significance test may be discussed in some detail and the student may even be set exercises in which he carries out such a test himself, but the aim of the work is to convey an appreciation of the purpose and logic of tests in general, and if this is achieved it would hardly matter if the specific details of the particular test were immediately forgotten. Similarly, introductory courses for other scientists might aim to provide enough background of ideas and terminology to enable the scientist to recognize when a problem arising in his own work has a statistical aspect, and to enable him to converse intelligently with a statistician about it, but without attempting to equip him with the detailed knowledge and skill required for its complete formulation, let alone solution.

In contrast, other students of statistics may wish to learn to use particular statistical techniques themselves without recourse to a statistician. The aim might be a limited one – like that of the beginning civil engineers who need to learn how to use least squares to 'correct' inconsistencies in surveying data – or there may be a broader desire to be able to use any of a battery of techniques already established in a particular field – trend-surface analysis, regression methods, analysis of covariance, and tests for circular data, for example, in geomorphology. Many courses will have aims which combine appreciation with a certain degree of technical competence: the student learns how to carry out certain fairly standard statistical analyses and, through the experience, comes to some appreciation of the wider possibilities of the subject. Service courses are often of this type. Such courses differ in *objective*, if at all, from courses for specialist statisticians or those studying statistics as part of a dual degree only in the emphasis given to the derivation of new techniques and the examination of their properties: the civil engineer or geomorphologist wishes to apply fairly standard statistical techniques to well-defined problems in his own field, though he probably does not hope to equip himself to derive methods or models for statistically novel situations, as the professional statistician might.

The aims of the student from another discipline who attends a service course in statistics overlap therefore with some of those of the specialist statistician: both seek understanding, and the statistician and some service-course students also wish to be able to use statistical techniques on substantive problems. A method of teaching desirable for specialists might also therefore be expected to be helpful for non-specialists: direct attempts to combine the teaching of statistical principles and methods with that of statistical practice will both deepen the understanding of principles and enable the student to use statistical methods more effectively. This is an argument for introducing some teaching of statistical practice into courses for all students, not just for specialists, whether the students' aim is to gain an appreciation of statistical ideas or to use them personally to a greater or lesser extent. Of course the teaching techniques appropriate for this will have to take account of the sometimes limited technical statistical background of the students and of the limited amount of time that they can devote to statistics, and may therefore not coincide with those used with more advanced groups. (See Chapter 3 for a discussion of teaching techniques, and Section 4.3.1 for some comments on the different requirements of elementary courses.)

Equally, for those who may wish to apply statistics in novel ways – which will mean in effect many of those studying it at a more advanced level, and particularly those studying it as part of a dual degree – a deliberate attempt to teach statistical practice alongside the systematized knowledge-based parts of the subject is as desirable as it is for specialists. Given the requirements of another subject or subjects, the time available for statistics as a whole and for practical statistics in particular will be less than it is for specialists, and so knowledge and skills cannot be expected to be developed so far. Two compensating advantages, however, at least for those studying statistics in parallel with a subject (such as economics or geography) other than mathematics or computing, may tend to offset this disadvantage: one is that work in the other subject may already have stimulated some of the skills, particularly of communication and evaluation, which a course in practical statistics tries to develop, and the second is that the student's perception of the role of statistics is likely to be clarified by his constantly seeing it in relation to the other subject. Indeed, in view of these factors it is not surprising that many graduates from dual-degree programmes have been so little disadvantaged in later statistical careers in comparison with statistical specialists. It is perhaps worth adding that the great variety of kinds of employment into which non-specialist as well as specialist students of statistics go makes all the more desirable an education which emphasizes transferable abilities.

Finally there is a small but important group of students who decide at an early stage on their future career and who study statistics together with another subject – often mathematics – to an advanced level, not with the aim of applying statistical ideas themselves in novel ways but rather as a preparation for research in theoretical statistics or for teaching. It seems equally important that this group too should receive an education in which statistical practice figures prominently. The reason, of course, is that a complete understanding of the subject is aided by such an education, and a complete understanding is particularly desirable in those who are to hand on the subject to future generations and who might in part influence its future direction.

1.6 PRACTICAL STATISTICS AND THE USE OF COMPUTERS

Computers have revolutionized statistical practice. They make large-scale data manipulation easy, they take the drudgery out of the application of standard techniques, and they have stimulated the development of new techniques, such as non-parametric regression, simulation testing and bootstrapping, which would have been unthinkable without them. As they figure so largely in current practice, it is important to ask what their place should be in statistical education, especially in that part of it directed toward practice.

Since it is obvious that long before the end of the working life of any statistician still being trained, or of one who has yet to begin his training, access to computers will be available everywhere – in the office, in the home, in the hotel room, and probably on trains and planes – it makes sense to have students using computers as soon as possible, and particularly using them in courses on practical statistics. The advantage of doing this is not only the general one just remarked: by reducing the time-consuming and tedious aspect of number manipulation and thereby allowing the student to concentrate on the important questions; it also makes better, more realistic, exercises practicable within his available time.

A point often made is that (professional, high-grade) statisticians should not just push buttons (or press keys, rather) without understanding what the machine is doing. With this no one could disagree, but occasionally teachers go on to argue that therefore one should not use computers in teaching, and in particular in practical classes. In our view this is quite mistaken: students should indeed be given drill exercises in, for example, analysis of variance to be done by hand, but not in the context of a course in practical statistics, which is trying to simulate reality and is concerned with quite different skills. The argument seems to us analogous to suggesting that logarithms should not be used in physics for fear that the students will not understand multiplication.

Granted that the student should understand what the computer is doing, what exactly does this mean? Clearly it is not necessary to know anything about the electronics; and equally clearly it *is* necessary to know, for example, which equations are solved in linear regression, but this leaves considerable room for doubt. Is it necessary to know *how* the equations are solved? Individual teachers will no doubt have their own views on precise requirements. Our opinion is that some attention to errors and rounding and other aspects of numerical analysis, such as methods for solving linear equations, is necessary. Knowledge of the general working of the computer, and of what a compiler, a high-level language, a package, etc., are, is beneficial, though not essential.

How should the computer be used? Almost anything one writes in any detail about computing is likely to be out of date before the ink is dry, so only fairly general points seem worth making. The extent of use depends of course on the resources available, but, assuming that they impose no constraints, students might well begin using the computer at a very early stage to cross-tabulate data, and to construct various forms of graphical presentation so that they may get the 'feel' of the data. They should progress as soon as possible to a state described by Pridmore (1985) as that of being

> 'keyboard-trained': that is, a state in which 'it will be second nature [for the student] to sign in to a computer system (just as easily as you or I get cash from a bank Cashpoint), pick out an appropriate menu and get down to calling up [a package] for a look at the set of data the lecturer has put on line for him to explore'.

They will then be able to concentrate on the more demanding aspects of statistical practice. There is little doubt that statistical analysis is almost always better done interactively (conversationally) if possible, because it saves so much time. But whether or not this is possible, it is clear that the ability to carry out further investigations on the results from a first analysis (for example to be able to regress the residuals from regression on some other variable) is highly desirable if not essential.

Two other questions are often discussed: is it better to use mainframe or microcomputers, if there is a choice, and should students learn to program rather than rely on packages? As far as the first is concerned, there is so much more and better software available at the moment for mainframe machines and their file-management capabilities are so much greater that they appear in our view to be the better solution, although the colour graphics available on most microcomputers but not widely available on the bigger machines would be useful in some contexts; but in any case the distinction seems likely to disappear as it becomes easier to link microcomputers to others. As to whether students should learn to program, presumably in a moderately high-level language like Fortran or Pascal, the answer is almost certainly yes – not all analyses will be directly available in a package, the statistician is not likely always and everywhere to have programming help available, and an ability to program, even if not often called upon directly, will increase confidence and give understanding of the computer's power and limitations. A practical point is that the language learnt should be one which will interface with the packages available. But the main concentration should be on packages, and the easier they are to use the better.

The foregoing discussion reflects a strong belief in the central importance of computing in the teaching of practical statistics, a belief which many teachers share. Occasionally, however, the belief is so strongly held that it leads to the confusing of statistical practice as a whole with computing, and the view that all that is necessary for acquiring practical statistical skills is the attainment of a certain dexterity in the use of the computer on statistical data. We believe that this view is as much mistaken as that described earlier which argues at the other extreme that computers should not be used at all in teaching. Computer-dexterity is certainly a vital part of practical ability in statistics, and it is a powerful aid to increasing that ability, but, as the discussion in Section 1.2 and in Chapter 2 illustrates, it is far from being the only skill needed in statistical practice.

1.7 BIBLIOGRAPHICAL NOTES

Although there are many books on statistical techniques and a few on statistical practice there are none that we know of specifically on the teaching of practical statistics. The conference proceedings edited by Rustagi and Wolfe (1982) discuss statistical consulting and how it may be taught. The book by Cox and Snell (1981) contains much wise advice about statistical practice, together with useful information on particular methods. On the other hand, many eminent statisticians have commented on the deficiencies in current statistical education and have suggested improvements. Besides the references in the text, several Presidential addresses to the Royal Statistical Society and to the American Statistical Association among others have mentioned these matters: see Bradley (1982), Finney (1974) and Yates (1968).

CHAPTER 2

Aims in the teaching of statistics

2.1 INTRODUCTION

In the previous chapter we have made the point that more is required of a successful statistician than just that he should be well-versed in theory, both because problems that arise in the real world are untidy and imprecise, in sharp contrast to the often elegant but inevitably simplified theory, and because, while any body of statistical theory is almost inevitably deductive in character, the making of a link between theory and practice essentially involves induction. In this chapter we shall expand on what exactly might be expected of a statistician, listing, and where necessary discussing, the various abilities that one might hope a qualified statistician would have. Before moving on to this, however, let us repeat, and indeed emphasize, that although we shall make almost no further comment about the theoretical background – the general principles of inference and probability theory, the detailed studies of particular methods, or the search for new or better methods – this silence is not to be interpreted as a suggestion that theory is not important: on the contrary, it is only through theory that we have any hope of reducing a vast mass of material to a coherent, manageable corpus of knowledge, and of understanding whether and why a particular method is valid or relevant in a specific context. Even the most applied of applied statisticians, indeed, surely needs the support of a solid theoretical background; the 'cookbook' approach, of learning methods or formulae to apply in cases which appear to satisfy certain criteria, is not enough.

It is perhaps worth making here the point that in an activity involving statistical ideas there are two characters, in the dramatic sense, whom for convenience we may call the investigator and the statistician: the investigator poses the problem in the substantive area – economics, for example, or industrial chemistry, or cryptogamic botany – while the statistician is skilled at manipulating and interpreting quantitative data. They may in fact be the same person, but if they are not then neither can give of his best unless a genuine partnership can be achieved by some means, if possible beginning at the early stages of planning the investigation. Here we are of course concerned only with the contribution that can reasonably be expected of the *statistician*. There is one general ability that a good statistician ought to have that we do not list – psychological insight (into how best to make this partnership fruitful); we have not said anything more about this since we very much doubt whether it can be taught or developed in an ordinary teaching environment, though it is true that some success has been reported in this direction from the USA in a rather specialized context – for small numbers of graduate students, with a substantial investment in material resources such as closed-circuit television and video recorders (Zahn, 1982a, b; Boen and Zahn 1982). It is very closely connected with the ability described in Section 2.2.1a, below.

2.2 DETAILED BREAKDOWN OF AIMS

The general purpose of courses on practical statistics is to enhance a number of abilities in the trainee statistician. The following list of desirable abilities, which is divided for convenience into four subsets, is, like any list, probably incomplete and certainly open to debate, but provides a convenient point of reference. It is of interest to note in passing that, although similar lists have been discussed in relation to a number of subjects, including mathematics, none directed specifically at *statistics* seems previously to have been published. (The discussion by Watts (1981) reflects a similar concern with broad objectives, but is restricted to a particular area of statistical method.) Two additional aims, though important, are not reflected in the list: to persuade students of the relevance of statistics, and to give them confidence to begin to apply what they in principle already know, to real problems.

Any statistician who had all the following abilities to a marked degree at the age of, say, 21 or so would be a paragon, so that the fact that the list appears formidable need not cause too much concern. On the other hand, anyone who did not have at least some degree of most of them, with an apparent potential for further developing the ones he has and for acquiring the others also, would not be an ideal candidate for a post of professional statistician. The list provides a framework within which to work, and against which to check courses or individuals.

Like most higher-level abilities, those on the list mostly do not occur on a simple presence or absence basis. For example, 2.2.3b – to apply any technique necessary – could apply to few if any in this world if *any* technique is to be interpreted literally.

2.2.1 General, non-technical, abilities

(a) **To work successfully with others.**

At the risk of stating the obvious, it is worth pointing out that there are many aspects of 'working successfully with others'. As well as purely social skills such as tact, it includes the ability in appropriate circumstances to take the lead or to follow another's lead, recognizing which kind of knowledge (statistical or non-statistical) is particularly relevant; the willingness to take on an appropriate share (either in subject matter or in time) of the total work-load; and an interest, both felt and expressed, in the outcome of the investigation.

Just as the statistician needs the investigator, so the investigator needs the statistician: a form of symbiosis. While it is debatable whether the statistician *qua* statistician can or should ever take the decision or draw the final conclusion, he can and certainly should be involved up to the final stage if at all possible: changing some of the conditions from those prevailing during the initial phase may well change the conclusions to be drawn, and a statistician is trained to look for such possibilities. This, like much of what we say, is describing the ideal, and for many reasons may not be practicable; but it is certainly less than ideal for either the investigator or the statistician to suppose that, by providing the figures, or by calculating an estimate, the statistician's contribution is complete.

(b) **To work to a time deadline.**

(c) **To communicate clearly and efficiently, both in writing and orally.**

One aspect of efficiency is, it may be noted, conciseness.

(d) **To appreciate the fact that in real life an answer must (usually) be found, however imperfect.**

That decisions must be taken is undeniable and a refusal to proceed because, for example, one cannot be quite convincing about normality, is unacceptable in the real world, but this aim needs to be linked with 2.2.2i (and also 2.2.3a and 2.2.4a): an essentially arbitrary decision should not be given spurious support by incomplete or only partially relevant statistical analysis. (Not all situations require decisions, of course, and in other cases some caution or hesitation may be appropriate.)

In the other direction, there may be more than one practically acceptable solution, and in that case the statistician needs to consider, or help in the consideration of, the various advantages and disadvantages, both statistical and non-statistical, of each.

One would be more hesitant about putting forward these rather obvious requirements, if one were not aware of how often they seem to be lacking in students – their inability to communicate in plain language is notorious, for example.

Some more detailed comments on how to achieve (b) and (c) are given in Section 4.4.

2.2.2 General, partly technical, abilities

(a) **To appreciate the ethical position of a statistician.**

A code of ethics for statisticians has been published by the International Statistical Institute (ISI, 1986), and this covers a wide range of issues, of which a number are worth mentioning specifically here.

The blatant falsifying of data or of arguments and conclusions is presumably no problem, since ordinary integrity will protect against it, but selective reporting of conclusions, or re-working the same data set looking for significant effects, are greater threats, which could arise either from conscious bias in, for example, an employer, or from subconscious pressures of enthusiasm on an investigator. Of course it is perfectly reasonable to re-analyse data using a better technique, or with a different set of questions in mind: the point is that any re-working changes the effective significance level, making it more likely that, even if no real effect exists, by chance one will appear to exist.

It is necessary to recollect that the data set does not belong to the statistician in most circumstances: it 'belongs' to the investigator, and the statistician has no right to pass it on to a third person or otherwise to publish it without consent. A somewhat similar point will often be relevant for the investigator rather than for the statistician – though there presumably might come a point at which the statistician should refuse to collaborate: the data are collected at some cost, and this cost may be measured in money, time, inconvenience, loss of privacy, loss of dignity, or even pain to some other party or parties, and it is essential to weigh carefully whether the benefits of collecting and analysing them justify the costs. Interestingly, this has some relevance in the context of teaching: a sample survey to be carried out simply as a learning exercise in effect usually makes the statistician coincide with the investigator, and a duty therefore exists to minimize the costs to the subjects, among others, and indeed to consider whether they, and the survey, can be justified. (If, of course, a survey topic with intrinsic value can be found, then the benefits are correspondingly greater, even if the costs are unchanged.)

This kind of consideration is especially important in certain areas such as market research and medical research, and in some areas codes of ethics have been drawn up which certainly have some relevance for the statistician.

(b) **To determine the aims of an investigation, and its framework, e.g. how the data are defined and collected.** (See also 2.2.4c.)

Here, as is the case with various other activities in these lists, part of the skill lies in disentangling the relevant points from a possibly very large amount of detail which adds little or no useful information.

This, and the next desirable ability, could normally only be properly exercised in conjunction with the investigator.

(c) **To translate general aims into reasonably specific and realistic problems, e.g. to decide that some factors are relevant and others not.**

In the sense intended here (see 2.2.4e for a more technical, but related, ability) this can only be exercised in conjunction with the investigator; indeed with an ideal investigator this might be unnecessary. Suppose, for example, that the investigator poses the question of forecasting sales of lawnmowers. It is possible to imagine a very elaborate enquiry relating sales to Gross Domestic Product, to the weather, to inflation rates and wage rates, and so on: but if one can agree that short-term forecasting, up to three months ahead, is all that matters, and that no substantial change in product or in the competition is involved, a relatively simple method based on the past sales may be quite adequate.

(d) **To recognize situations which call for checks or controls on the quality of data, and to construct suitable procedures for carrying out such checks.**

Large-scale investigations, or those spread out in time or space, are obvious cases requiring this treatment. Almost any survey involving more than one interviewer would need some attention under this heading to ensure consistency, completeness, and so on.

(e) **To organize work, e.g. data collection and data analysis, effectively.**

This might almost have gone into section 2.2.1.

(f) **To recognize the limitations of one's knowledge.**

It will be difficult to achieve this, and the following aim, unless one is scientifically – in the broad sense – literate and aware.

(g) **To find and read, critically, other relevant material both in statistics and in the subject area of the investigation.** (See also 2.2.3d and 2.2.3e.)

This should imply an acquaintance with general bibliographic aids, such as abstracting and indexing journals, at least in statistics.

A statistician working more or less permanently in a particular situation, e.g. for a drug company, is likely to acquire quite rapidly a good deal of general background information that will be relevant to what he does (though it will take him a long time to become a real expert); at any rate, if he does not he is likely to be thought not much of an asset. A statistician working in a less specific environment will have more difficulty in reaching such a state, and will need to develop as far as he can the art of picking up the essentials of the context quickly.

(h) **To interpret and/or utilize the results of the analysis.**

This point, which links up with 2.2.1c and 2.2.3b, and the next one should perhaps be considered together: the basic point here is the familiar one that the answer to a practical question cannot be 'the answer is significant'. One important aspect, often neglected, is the question of to which population do the results apply: more generally, how widely do the results apply.

(i) **To understand what statistics can and cannot do.**

Perhaps this might be paraphrased as 'to stand back from the data'. To recognize when extra-statistical input such as a value judgement is being made, and that uncertainties over exact values of losses in many contexts cannot be resolved by ever more sophisticated statistical analyses, are typical examples of this; cost-benefit analyses, in which the sheer weight of data available for some costs hides the fact that other costs are not really measurable, are not uncommon – studies of the need for a new airport, for example, in which travellers' delay-times have to be balanced against the loss of a community, or of a unique mediaeval building.

2.2.3 Abilities depending on technical skills

(a) **To recognize which techniques are valid and/or appropriate.**

It is this aim which necessitates even very applied statisticians having a good working background of theory; it may, of course, be the case that no existing technique is obviously valid or appropriate, and this too should be recognized.

It is important, in particular, to be aware of the various kinds of measurement – to distinguish counts from continuous variables and so on – since some techniques are inappropriate for some kinds of data: the use of χ^2-tests with count data measured in thousands, or with percentages, is by no means unknown, for example.

Indeed χ^2-tests are associated with several particularly frequent errors: testing for independence when it is hardly to be doubted that dependence is present and something like a measure of association is needed instead; testing for independence or homogeneity on data which in fact make up the whole population. As far as the former point is concerned, the fact is, of course, that in the great majority of situations estimation is more appropriate than testing.

(b) **To apply any technique(s) necessary (and to interpret the results and draw valid conclusions): comprehension.**

(c) **To apply any technique(s) necessary: computing and similar skills.**

For some comment on computing, see Section 1.6.

(d) **To find and use the main sources of published data.**

Although this does not now appear in many statistics courses, it seems to us that the ability to make good use of these sources is a real technical, although not theoretical, skill. Both this and the following point are extensions of 2.2.2g.

(e) **To learn to understand and use a previously unfamiliar technique.**

If, for example, it becomes clear that cluster analysis may be relevant, an ideal statistician will be able to go away, learn about it, and come back and apply it.

Since no course could possibly cover all known techniques in sufficient depth for them to be used – let alone those that will not be invented until after the end of the course – this is perhaps the most crucial ability of all: it is the ability to teach oneself, or to go on learning. All full-scale *educational* courses – in contrast perhaps to training for particular purposes – would acknowledge this as a central aim.

Using such an ability is, however, another matter entirely. No one is likely to have the time to master a new field of any size very often, and it becomes important to judge when it is appropriate to call in an expert (2.2.2f), or to conclude that insufficient gains will come from this new technique (2.2.4b).

2.2.4 Abilities depending on technical judgement

(a) **To appreciate that real data will have imperfections, and to react sensibly to difficulties.**

A simple illustration is provided by sample survey data in which the sampling frame is out of date – as it almost always will be, at least to some extent; again, it is sometimes clear that a slight dependence is, or may be, present; as another example, trade and production statistics are often collected using somewhat different classifications. A statistician who completely ignored the problems would soon lose his reputation, while one who refused to analyse data unless they were known to be perfect would be of little value. The right reaction will vary from occasion to occasion and will depend on the extent of the difficulty and the importance of the context in relation to the available resources: sometimes it will be clear (i.e. one's judgement and experience tell one) that there is no problem at all; sometimes the difficulty will leave largely unquantifiable uncertainty in the conclusions, although the qualitative conclusions can hardly be doubted; sometimes a great deal of effort will need to be expended to overcome the worst of the difficulties; and sometimes it really will be appropriate to refuse to go further. This is closely connected with the following.

(b) **To recognize the various levels of sophistication of techniques of analysis which are appropriate for data of different reliability and importance.**

If the data are not recorded to any great degree of accuracy, and only relate to part of the problem (cf. 2.2.2i), it is unlikely to be worth while to pursue an analysis of great sophistication; optimal – and, even more, asymptotically optimal – procedures often add little to rather routine approaches (though of course they provide a standard against which to measure the efficiency of other approaches), and are, consequently, often scarcely worth worrying about. Not using all the information (by, for example, converting a scale of quality into an acceptable/unacceptable dichotomy) is, on the other hand, not to be recommended unless one has some reason to believe that no serious loss will ensue.

In particular, excessively precise calculation of statistics or of quantities such as significance levels are to be avoided; if, as is commonly the case, a computer package produces unreasonably precise results, they should be appropriately rounded before being reported.

(c) **To choose, or develop, an appropriate plan or design for an investigation.**

The appropriateness or otherwise will certainly depend on how the costs compare with the benefits to be obtained: the immediate purpose is to gather information while making good use of limited resources of time or money. It is at this stage that the statistician can often justify his existence, for no analysis can be worth more than the data on which it is based, and an inappropriate design may mean that no worthwhile data can be collected. The 'design' in the conventional statistical sense is not the end of the story by any means: even if it were agreed that, say, a latin square is the best design in some context, the questions of exactly what is to be measured, and how, by whom, and to what accuracy remain, as does the choice of the means by which the data are transferred from the recorder (instrument or person) to the statistician. If a statistician is to earn his keep, he has to know what assumptions about the data are reasonably made – what model is valid, for short – for it is this which determines whether or not the later analysis is justifiable. Any statistician who has worked with other investigators will have a fund of stories about data which were not what they were claimed to be: contaminated, missing but filled in (arbitrarily) to complete the table by a technician, dependent because the measuring instrument was not cleaned properly or reset after each measurement, and so on.

A subsidiary aspect is concerned with the ability to write project proposals.

(d) **To assemble seemingly unrelated facts into a pattern.**

(e) **To translate a real problem into a statistical form.**

This naturally follows 2.2.2c, and is linked with the next ability. The needs here are various: it may be a question of deciding that the right formulation is one of discrimination rather than of hypothesis testing, or of recognizing that it is a prediction interval for a variable rather than a confidence interval for a parameter that is required. Or it may be at a somewhat more fundamental level, such as a decision about whether it is useful to think of a particular (real) problem in terms of point processes, or density estimation, and so on, and, if it is, what would be the various components of such a formulation.

(f) **To build models.**

The distinction between 'explanatory' and 'pragmatic' models (Cox and Snell, 1981), i.e. between models which at least try to copy the microstructure of the situation and those which are merely descriptive, is important. A model of the latter kind, for example a simple linear regression, may be quite enough, and may in any case be the best that can be done, since the microstructure may not be understood; but one would have less confidence in its applicability in different conditions.

Too much stress should not, however, be placed on the distinction, since these represent only opposite ends of a continuous spectrum: any model is only a partial representation of reality and it is thus in the end a question of to what extent a particular model copies reality, and how far it has been tested against reality.

(g) **To develop new methods.**

2.3 DISCUSSION

These four subdivisions of the abilities that one might hope a statistician has by the time he completes his formal education correspond, very roughly, to different levels of *statistical* achievement: anyone, statistician or not, should meet the requirements of Section 2.2.1, while Section 2.2.2 corresponds to statistical literacy; Section 2.2.3 might be asked of a statistical technician, while the final group, in Section 2.2.4, is more appropriate for a fully qualified professional statistician. In particular, for courses with clearly limited intentions, such as service courses for other disciplines, it would probably be sensible to regard Sections 2.2.1 and 2.2.2 as describing the aims. But while it may be correct to relate the four lists to differing levels of *statistical* achievement, it certainly is not the case that in any other sense the order of the lists necessarily represents increasing maturity and experience or intellectual ability: 2.2.1.c, 2.2.2.e and 2.2.2.g, for example, are by no means trivial skills.

How far one or two of these requirements can be taught, in the narrow sense, is a moot point – for example, 2.2.1a, 'To work successfully with others'; but at least attention can be drawn to them and some opportunity provided to try. Some others are difficult to provide for in the confines of a course – for example, 2.2.2d and 2.2.4d – but the essential point is that the nearer one comes to developing in a student at least the rudiments of those abilities, the better a statistician he is likely to be. As noted in Section 1.4, it has sometimes been argued that it is not necessary to concern oneself with these questions, and that common sense and intelligence are all that is needed. We do not think this is the case, since even if it is possible in principle it is certainly inefficient to make all the mistakes and to learn all the lessons from bitter experience. As Henry Ford is alleged to have said: by the time a man is ready to graduate from the University of Life he is too old to work (to which George Bernard Shaw added: and the fees you have to pay in that school are exceptionally high). In the next chapter, therefore, we shall discuss in some detail the various arrangements that have been proposed for teaching practical statistics, and the extent to which they provide for the requirements listed in Sections 2.2.1 to 2.2.4.

CHAPTER 3

Approaches to teaching practical statistics – a survey

3.1 INTRODUCTION

Although there seem not to have been any substantial discussions in print of how the practice of statistics might be taught, nor of what precisely should be included in the term 'practice', every teacher, or teaching group, must have considered such problems, in general terms at least, and adopted a strategy to deal with them. The outcome has been a wide variety of approaches to bringing what is taught in statistics programmes into contact with what is done in practice, and the purpose of this chapter is to review them in a systematic way.

Although the rather detailed listing of aims that was given in Chapter 2 is useful in showing just how many different kinds of skill are needed by a statistician, and thereby underlining the fact that many traditional statistics courses omit all consideration of some of the most important aspects of the practice of the subject, it would be tiresome to work through the list for each of the various teaching methods discussed in this chapter, listing those which are or are not, or may or may not be, exercised and developed by each approach. In any case the success of many of the methods in fostering the various abilities depends critically on the exact details – which case study, or which project, for example. In our discussion, then, we point out the advantages and disadvantages of the various approaches in general terms only; a detailed analysis could be made, if it were thought important, in any particular case. One aspect, to which we make passing reference, is that of the assessment of students: many of the approaches lead to considerable difficulties in this respect, and yet a student who works hard at a particular piece of work deserves some reward. There seems no entirely satisfactory solution in most cases.

Probably few teachers think of any approach, even if it is one that they have themselves adopted, as ideal. Conversely, each approach is certainly valuable in its way. Some are more demanding of resources, or of a particular organizational framework, than others, but there is no doubt that if there were time enough for all of them, and for the basic theory, all would contribute to the development of valuable skills in the student.

That all are valuable is true , but all are not, presumably, equally valuable. We shall not attempt to balance one against the other, however, partly because we recognize the subjective nature of our opinions on the relative importance of the aims listed in the previous chapter, and partly because the relative merits depend, in some cases at least, on detailed characteristics of a particular group of students within a particular institution with particular goals. In keeping with this, the order in which the following sections appear has no especial significance.

3.2 DRILL EXERCISES

We mean by drill exercises numerical problems for which the technique of analysis is standard, and is either prescribed or is fairly obvious, and for which there is an understanding between the student and the teacher that a solution will be judged adequate if it is formally and numerically correct, though there may also be a requirement to make a gesture in the direction of translating conclusions into a verbal form. Exercises like this often accompany courses on statistical

EXHIBIT 1

Example 1 Twenty-five plots, of approximately equal fertility, were sown with 5 different varieties of wheat, 5 plots to each variety, the distribution of varieties among the plots being random. The yields of grain in bushels per acre are contained in the file YIELDS, the 5 columns corresponding to the different varieties. Investigate, using MINITAB [or other convenient statistical computing package] or otherwise, whether there are significant differences among the yields, and draft a report to the research station which produced the data.

Example 2 The table below gives the measurements in millimetres of lengths and widths of the second of the lower cheek teeth of 30 adult specimens from two subspecies of *Megalocnus rodens*, the giant sloth.

M.r. rodens		M.r. cassimbae	
Width	Length	Width	Length
23.8	16	21.4	15.2
23	15.5	16.5	14.8
.	.	.	.
.	.	.	.
.	.	.	.

Source: Matthews and De Conto, *Bull. Amer. Mus. Nat. Hist.* (1959)

A new specimen has second lower cheek tooth width 18.7 mm and length 15.2 mm. On the basis of these measurements, to which, if either, of the two subspecies would you allocate the new specimen? How would your analysis alter if the population proportions of the subspecies were known for the region in which the specimen had been caught?

methods. Those shown in Exhibit 1 might be examples, depending on the context in which they are used.

Against the background of a course on elementary statistical methods Example 1 in Exhibit 1 could be a useful means of teaching the arithmetical or computational operations needed to obtain a simple analysis of variance, and it might also have some value in consolidating the student's knowledge of the associated theory. For these purposes an 'adequate' answer might be made up of the output from the one-way analysis of variance instruction in MINITAB, together with a formal test of the varieties mean square. (It might be supplemented, if the student had enough knowledge, by a check on within-group variances and a comparison, if necessary, of group means.) Similarly, for Example 2 in Exhibit 1: in the context of a course on multivariate analysis, a routine application of a discriminant analysis procedure might suffice. Sometimes, in exercises such as these, students are given individual artificial data-sets and the teacher, knowing the corresponding right answers, is provided with an easy means of checking that each person separately has performed correctly.

Drill exercises are widely, probably universally, used, but, even today, there are some statistical programmes in which they provide the only means by which the student can meet numerical data. There can be no doubt of their value in illustrating and consolidating knowledge of the mechanical aspects of formal techniques – indeed they are almost essential for these purposes – but it would be a serious mistake to suppose that they adequately exercise the wider skills of practical statistics. They are to our subject what five-finger exercises are to piano-playing, or spelling bees to literary composition. The dangers are that able students will find them insufficiently challenging and that they will give students a distorted view of statistical practice. It seems trite to point out their limited value for fostering any of the skills of synthesis, evaluation, and communication central to the practice of statistics, and we would feel embarrassed at doing so if it were not for the fact that numerical drill exercises have been so widely confused, in teachers' minds and in the construction of statistical programmes, with the essence of practical statistics. Their overriding importance is in developing skills related directly to knowledge of techniques – 2.2.3b and 2.2.3c in particular.

It is, of course, possible to make exercises such as those above more demanding. If the substantive background to the varieties trial were given more emphasis, for example, and it were made clear that an answer in terms of the agricultural question was required, with attention being given to the validity of any formal statistical method used, then the numerical or computational part of the exercise would no longer dominate, and the student's judgement – about whether assumptions are justified, whether a transformation is desirable, whether information about other aspects of the experiment or other data should be sought – and his powers of interpretation and presentation would be much more fully called upon. Even though the main technique used – analysis of variance – is not in question, so that skills associated with formulation of a problem and choice of technique are not exercised, the benefit would be substantial. The guidance required by the student from the teacher, of course, would increase too. Equally, in other contexts Example 1 of Exhibit 1 *as it stands* could demand a great deal more than routine numerical processing: if students had never met analysis of variance, for example, it would be quite challenging. Changes to drill exercises of this kind transform them into one particular type of 'mini-project' (perhaps especially suitable for less advanced students) of the kind discussed in Section 3.14, and more fully in Chapter 4.

In recent years a rather different-seeming type of exercise has become popular in some quarters, with the ready availability of computer packages. The data-sets put forward for analysis have become rather bigger, and the techniques to be applied more numerous – not just the fitting of regression surfaces but also the consideration of C_p plots, for example. On occasion the task set may be to disentangle the complexities of a package-based set of analyses, presented in the form of computer output, with a view to interpretation. For all their differences in detail, however, such problems seem best described as drill exercises.

As far as the ordinary kind of drill exercises is concerned teachers often tend to be rather ambivalent. On the one hand they recognize that the aim is a limited one, and implicitly admit this by using artificial data, and on the other hand they feel the need to dress the problems up with circumstantial detail, as if to say that the data *might* have arisen like this, even if it didn't really. It seems to us that drill exercises can, and probably should, be limited in scope and size, and not provided with practical decorations which only distract, and they should not be assumed to provide all that is needed to train students in the skills of practice. The virtues in the provision of at least a little practical decoration are that it allows some practice in the translation between formal statistical language and everyday language, and it helps to illustrate the variety of problems which can be tackled by statistical means. The observation by Searle (1971, preface), following Li (1964), that just as one does not learn to solve quadratic equations by studying equations such as $683125x^2 + 1268.4071x - 213.69825 = 0$, so there is no reason not to have carefully chosen numbers, to make the arithmetic simple, in statistics exercises, is relevant here, in showing by analogy both the value of drill exercises and their limitations.

3.3 STATISTICAL EXPERIMENTS

A number of writers (for example Jowett and Davies, 1960; Scòtt, 1976) have described the use of experiments in the teaching of statistics. What is meant by an 'experiment' varies considerably, from the use of random number tables to demonstrate the central limit theorem, through exercises which are artificial but involve real data, such as measuring diameters of large numbers of ball-bearings, to small-scale but genuine enquiries, such as traffic surveys. The name 'experiment' is in some cases rather misleading, and not all authors use it, but it is a convenient short term to use in the present context. Normally students are given quite explicit instructions about what to do and how to do it, so that one might consider these experiments to be generally rather similar to drill exercises as described in Section 3.2. The benefits to the student from such experiments are variety of interest, and the requirement of a rather more active role in relation to data collection, so that at least some appreciation of what can be an extremely difficult task is gained. Those experiments which are not artificial have the additional advantage that unforeseen difficulties may present themselves (for example, some data may be missing, or there may be problems of definition – is a Land Rover a car or not?), and students will then at least encounter such difficulties, and may even have to make decisions about what to do. Often there has been a requirement to report both orally and in writing on the results.

Most such experiments will be time-consuming at the data-collection stage even though data collection is usually made a group activity, and the more obviously artificial ones are likely to be found uninteresting by students, particularly those in older age-groups, though schoolchildren may not be quite so critical; but such tasks can certainly be valuable, if interest can be engaged, in exercising some of the abilities looked for, such as those related to communication, and in encouraging cooperation in groups – indeed, as we point out in Section 4.5, we believe that a well-chosen experiment provides opportunities that few other activities do for developing such abilities.

When experiments are used to complement some other course – to illustrate the central limit theorem, for example – there seems no cause to try to assess them, but if they stand by themselves, as they are likely to in any attempt to teach statistical practice, the teacher may well feel obliged to attempt assessment. This certainly presents problems if the experiment is carried out as a group activity: participation might be measured, presumably on a very coarse scale, and clearly a final report could be graded, but there is no avoiding the fact that it is difficult in most cases to distinguish an individual's actual contribution, and impossible to discover what he might have contributed had someone else not said or done it first.

3.4 CRITICAL READING

Hawkes (1980) has described a final-year course in practical statistics based exclusively on the critical reading of statistical papers. Each week the class is given a paper to review from the applied statistical literature. The general aim is to understand the purpose of the paper and the methods it uses, and to evaluate and comment on the extent to which it is successful. Specific questions about details of content might also be assigned to help break down the large aim. It is suggested that, if necessary, a modest amount of re-working of the data in the paper might in some cases be carried out, but students are not asked to prepare a written report on their review. Instead they meet as a group with the teacher to discuss the paper and to give an oral account of any further analysis they have done.

Assessment for the course is by a written examination, in which specific questions about the detailed contents of a selection of the papers reviewed throughout the year are set (and copies of the papers are to hand). It is possible for a student to take very little active part in the weekly discussions, but the final examination provides a disincentive to total inactivity. Nevertheless, some students might be tempted to economize on their own efforts in the hope of picking up enough from their more industrious fellows at the weekly discussions to help them through the examination. With honest effort by participants, on the other hand, the discussion group can result in each member's gaining a well-balanced evaluation of the paper, and, more importantly, can provide an illustration of the way a rational evaluation is arrived at, and an appreciation of the criteria of evaluation. Weaker students in particular are likely to benefit from this, though the requirement to write a report of their own would, perhaps, consolidate the lesson. Hawkes gives a valuable list of papers suitable for use in this kind of course. With careful choice of material the class can see a variety of statistical methods in action in many different fields. Limitations are that, nevertheless, only certain types of problem (fully formulated, moderately scaled, suitable for publication) will be seen, and that skills called upon only when the statistician has to take full responsibility for analysing and reporting on a set of data will not be exercised: for example, working with others (2.2.1a), determining the aim of an investigation (2.2.2b) and organizing work (2.2.2e) will not be called on. Although many of the other abilities listed in Chapter 2 can be – though they may not be – *illustrated*, they will mostly not be exercised and developed by the student himself.

3.5 LARGE PROJECTS

Most, and perhaps all, postgraduate (MSc and MS) programmes based on coursework require students to present, as part of the evidence for the degree, a substantial dissertation. Such a requirement is much less common, though by no means unknown, in first-degree programmes; in this context it is usually thought of as the equivalent of one or two courses, contributing anything up to 25% of the total work-load and often of the assessment, and is frequently optional. Informally, presenting a dissertation is often described as the student carrying out a project (and then the dissertation is the project report), but regulations are in fact rarely explicit on what the subject matter of a dissertation should be; custom, frequently at departmental level, may constrain the subject matter to be within sometimes quite narrow limits.

There are wide variations in the details of the way in which such projects are arranged: the stage in the programme at which they shall be started and completed, the extent to which their progress shall be formally monitored, the source of the idea (assigned by staff; chosen by the student from a selection offered by the staff; proposed entirely by the student, though approved by staff, with a back-up system to provide subjects for any students who fail to produce one of their own), and to what extent help is available, for example. Moderately detailed descriptions of how projects are arranged in particular institutions are given by Griffiths and Evans (1976) and Kanji (1979).

The subjects that have been used for 'statistical' projects have also been extremely varied, ranging from purely theoretical and rather abstract surveys in probabilistic or statistical areas, through analysis of moderately substantial but fairly conventional data sets, to the complete design, execution, and analysis of a sample survey, with forays into numerical analysis, computer program writing, and the writing of critical reviews in areas such as operations research as well as in more traditional areas such as statistical inference; almost anything, in fact, that will not be too easy or too hard and will take about the right amount of effort.

The value of projects can be judged in a number of ways. Whatever the subject, and however bad a project might be thought by critics, the student will have had to devote considerable time to a single piece of work, to organize his thoughts, to plan at least to some extent the strategy of the investigation, and to write a moderately lengthy piece of prose, the whole to be completed by a prescribed, and probably inflexible, deadline, and most of these demands will occur in no other part of the programme. With a careful choice of the subject, and a certain amount of luck, a fairly substantial proportion of the abilities listed in Chapter 2 can be engaged, and in a way that is not otherwise really possible in teaching contexts except for external placements (Section 3.9) and consulting courses (Section 3.8). Clearly it is well worth considering whether a project, or indeed more than one, can be included within the programme.

But of course it might not work as well as that, and caution is needed before deciding that a project is the answer to the problem of teaching practical statistics. To find projects that will bring out all, or at any rate most, of the points that one wishes to emphasize, in sufficient numbers for a class of any size is very difficult, and it is in the nature of a large task such as a project of this type that one has to put all one's eggs in one basket – if after some time it becomes clear that the project is not really as good as it seemed, or just that it is not working properly because of unforeseen difficulties, it is usually too late to change things, and one may end up with a minor disaster. In some circumstances one may be able to find time for two or even more large projects, but this depends on the length of the programme and will be unusual, and in a rather different sense one is therefore usually again putting all one's eggs in one basket (unless one regards the project as merely providing the climax to a planned development of practical skills). If the project is the first and only occasion on which a student has to concern himself with working in a systematic way alone, and with writing in English, and so on, it is difficult for any feedback – for any teaching of these aspects, in other words – to take place. It is almost certainly, in fact, necessary for a project to be preceded by some other form of teaching about statistical practice.

As in many other situations in which students attempt tasks unique to themselves, assessment is difficult, and this is doubtless one of the reasons why there are typically more middle marks and fewer high or low marks in the grading of large projects than there perhaps ought to be.

There are two other points worth making. Firstly, there is another possible benefit to some students: the opportunity to carry out a project may allow a student to choose (or may by accident introduce him to) an area in which he will later wish to work or do research, when it would be quite impracticable to teach a specific course in that area. Secondly, a rather different type of project has been assigned in some programmes – group projects (one or more depending on the size of the class) in which a moderate number of students work together, for example to carry out a sample survey. Such a project, which can be more ambitious, has rather a different purpose and in consequence it may be inappropriate to assess it other than as satisfactory or unsatisfactory. Such a framework will, with appropriate choice of subject, *illustrate* many of the problems associated with practical work, and may do it very well because of the scale of work possible; but the larger the project, the more the tasks will need to be divided among those taking part, so that any single student may experience directly only a small proportion of them. The one skill that should be particularly exercised is, of course, 2.2.1a – working successfully with others.

3.6 APPLICATIONS OF STATISTICS IN ANOTHER SUBJECT

There are several ways in which another subject can in some sense be linked with statistics in a student's programme. It can be merely a question of the student simultaneously taking courses in statistics and in the other subject, and the two may then have roughly equal weight, or one may be clearly of secondary importance (a subsidiary subject); there is in this case little or no attempt at really linking the subjects. There may be an attempt to construct a joint programme (in, for example, statistics and economics), so that the two parts complement and supplement each other; this aim seems to be not always achieved, even when it is declared. Neither of these approaches has as its aim the enhancement of the ability to do practical statistics, although the need to develop rather different habits of thought may, particularly in a well thought out joint programme, contribute to this; not every such subject is treated from a practical point of view, in any case. A final category is of the kind 'Applications of statistics in (or to) medicine' – or in economics, agriculture, etc. If, as such a title suggests, the course consists to a large extent of case histories (Section 3.10) or case studies (Section 3.11) of various kinds it certainly provides a real-world context within which statistics can be seen to work, and it may in some cases go much further, for example in involving a student in the collection of data in ways which are of particular interest and relevance in the other subject. There is a possible disadvantage in such a course except for those going on to specialize in the area, in that so far as the training of statisticians is concerned the main aim is to assist in the development of skills which are transferable to other contexts: in many cases much of the time and effort devoted to mastering the concepts of the other subject will not contribute to this, though learning how to learn is a skill that may be enhanced. (There are of course other arguments to suggest that every statistician ought specifically to know some economics, and doubtless a number of other subjects can be argued for also.)

What is certainly of somewhat doubtful utility for fostering practical skills, except in a narrowly vocational sense for those committed to a particular career and except in so far as it motivates the students, is a course which merely discusses statistical methods of particular value in the other subject.

The benefits and disadvantages of the present approach are largely those of case histories and case studies in general which are discussed in Sections 3.10 and 3.11.

3.7 CONTACTS WITH OUTSIDE STATISTICIANS

In most educational institutions – universities, schools, and so on – there are few if any statisticians who spend a particularly large part of their time actually doing statistics, as opposed to teaching it; and even when one finds such a statistician, he tends to be a consultant, who almost by definition sees rather special kinds of problems, problems too which are already partly formulated in many cases. Thus the idea of bringing students into contact with working statisticians from outside the academic sphere is an attractive one.

There are basically two ways of doing this – bring the statistician to the students, or take the students to the statistician; the former is usually easier, though with careful planning the latter may provide more benefit – it may then be possible to see, vividly, the office routine and the organization and structure within which the statistician operates, to meet associated specialists such as computer professionals or even just a wider variety of statisticians, and also of course to see the raw material from which the data arises, whether they are fields of sugar beet or production lines for deep-freezing peas. In either case, in so far as the main activity is oral, the working statistician can talk about these other aspects, but is likely to spend much of the time presenting case histories (Section 3.10). The student will thus hear of many of the important features of practice, but of course he will not in this way be able to meet them face to face or to demonstrate his own ingenuity, toughness of mind or systematic working habits. Indeed, unless the meetings with statisticians are developed into a rather formal activity – such as an organized sequence of case histories, with opportunities to discuss them, at the least – they are likely to seem rather lightweight in nature, offering little more than motivational/inspirational talks; and any idea of assessing the students would surely be inappropriate.

If a visit by students is to take place, advance preparation, so that they know why they are going, and, at least in general terms, what to expect, and how the various aspects are related, is essential; not all working statisticians know intuitively what is expected of them in such a context – it is not clear why they should be expected to – and preparation on this side may also pay substantial dividends.

3.8 CONSULTING COURSES

Many larger teaching groups have a second function – to provide a statistical advisory service to others who need it; sometimes it may not be the teachers themselves, but associated consulting staff who provide the service. (The detailed arrangements, determining, for example, who may use this service – other academic groups on the campus; local hospitals and government in addition; anyone – and on what basis (for payment or not) vary considerably but do not matter greatly in the present context.) When such a statistical advisory service does exist, it may be possible to use it to provide valuable training for students. If resources – mainly student and staff time – were unlimited one could envisage a steady development in which at the beginning students merely listen to and watch an otherwise ordinary consultation; at a slightly later stage a discussion, either individually or in groups, on the way the consultation proceeded, would be valuable; a third stage would be one in which the student played an active, though clearly junior, role in a consultation; and the final stage would have the student being in effect the consultant, with all the responsibility that entails from first meeting the client in order to begin to disentangle the problem through to preparing a final report – he might even be able to negotiate any contract needed – with the staff being involved only in order to ensure that no disaster occurred. Some groups have been experimenting with this approach for a considerable time and find that it works well; on the resource side it has been extended so that video-recordings of successful consulting sessions are available for study by the students, and so that recordings can be made of sessions in which students are actively involved, which allows a convenient and easy review of the progress of the consultation. In this case interpersonal qualities can also be discussed and perhaps taught – tactful persistence in eliciting the real problem, making people feel at ease, and so on. (See Boen, 1982; Zahn, 1982a, b.)

As an approach to training fully professional statisticians a consulting course is very attractive, since if all these things can be done, and done well, with a reasonable variety of consulting problems, then of course one has largely overcome the difficulties in turning students into statisticians: virtually all the abilities listed in Chapter 2 will be called on. But it seems unlikely that it will ever be even part of the solution for most students, and perhaps never the whole solution for any students. What, after all, does it require? Students must, at least if they are to go on to the later stages, be already trained widely in the theory and technicalities of statistics, so the approach can really be used only for those training to be full professionals; moreover, the pace at which it is supposed to take students to the stage of accepting effectively full responsibility is probably too rapid for more than a small proportion of students. Even if the desirability of physical resources such as video-cameras and recorders is agreed, and the funds necessary are available, the staff time required is substantial, and, in contrast with most other approaches, almost directly proportional to the number of students involved, making it impracticable in many cases. More, student assessment is likely to be difficult. It should be noted, too, that advisory services do not necessarily attract a wide variety of *types* of problem, and the implications of this would need to be assessed carefully. The subject matter areas which present themselves (biology, industrial quality control, etc.) may well be limited, though this is probably not very important, unless the range of problems is also limited; it is more serious if the problems are usually moderately well formulated, so that the advice required is more nearly technical in character. It is clearly important to realize that acting as a consultant is only one of the possible modes that a working statistician may be in.

Nevertheless, a consulting course (or an apprenticeship or internship scheme), if it can be made workable, is potentially very valuable and can certainly be used to supplement other approaches. It can serve, incidentally, a rather different purpose very well: by showing the other aspects of statistics, such as the non-textbook nature of typical problems, it can provide motivation and a context for further study.

Several of the articles in Rustagi and Wolfe (1982) discuss the teaching of consulting, and there is a somewhat earlier bibliography on statistical consulting (Woodward and Schucany, 1977).

3.9 EXTERNAL PLACEMENTS

A number of institutions have arrangements whereby students spend a period working in a statistical environment: in a government office, or in an industrial quality-control department, for example. Such arrangements go under various names, depending to some extent on the details, and on the customs that have built up: sandwich courses (with 'thick' or 'thin' sandwiches according to the length and number of the placements); cooperative programmes; (summer) placements; professional experience; and in the French tradition 'stages'.

The precise intention of such a placement, the means of accomplishing the intention, and the way in which the progress of the student is monitored while on the placement are again variable, but broadly we may say that the aim is to develop practical skills in the student by placing him in a real working environment for an extended period, and to motivate him to further study.

If time and money were no problem and if one could find suitable locations to send the students to, one might imagine that placements of (a minimum of)

3 months after a first-year course (i.e. a first course of perhaps 100 lectures) followed by (a minimum of) 6 months after the second year (another 100 hours or more), with both carefully tailored to the student's experience, so as to use his knowledge and provide a challenge at the same time, with a regular drawing together by a supervisor of what was being learnt, would make it possible for all the potential advantages to be realized. Then the student would be taught to see how all the various parts of the subject relate to their applications, and would get the benefit from, and the satisfaction of, being involved in the complete process from initial planning to final reporting of a project; all, or at least the very great majority of the skills listed in Chapter 2 would have been exercised and to some extent developed. Further advantages would follow: the student would know whether this kind of work appealed to him, and in at least some cases both employer and student would agree that permanent employment after completion of the course would be an attractive idea.

To what extent these advantages are realized no doubt varies greatly. It is very easy to see how things might go wrong: an employer, through ignorance, ill-will, or just misfortune, might use such a student – admittedly only partly trained – as just an extra pair of hands, and certainly the employer's main aim will not be the same as that of the institution. We have known of supposedly statistical placements in which the student spent much of the time in one case writing a computer program, and in another managing a database – both no doubt very valuable from some points of view, but not what one has in mind in recommending placements; in an extreme case it seemed, from the description given by the student, that she was acting as little more than a filing clerk. Careful management and supervision of the kind envisaged above is also perhaps not as common as one would wish, and in few placements will it be possible to provide an especially wide variety of experience. The problem of supervision, and of assessment if this is thought appropriate, will be considerable. It is certainly possible for a placement period to be an almost total waste.

But in fact the choice of whether to use placements is more likely to be made on grounds less specific to statistics: the institution's rules and customs may either not allow it at all, or almost require it. It is likely to be appropriate only at degree level in any case. Arranging placements can absorb a great deal of staff time, and it is also necessary to reflect on whether delaying graduation for a year, which is the usual consequence of a placement scheme, is really worth while. For a general study (not related to statistics specifically), see DES (1985), and for a description of the experiences and practice of one group, see Jones and Kanji (1980).

The one overwhelming advantage that placements have in comparison with all other methods of linking theory and practice is that they do not involve role-playing, and of course to see a working environment from the inside is almost certain to be educational to the student, at least in a broad sense. Certain kinds of things, such as office practice, and personal relationships in a working environment, are perhaps best learnt this way.

3.10 CASE HISTORIES

A case history is a narrative account of the investigation of a particular real problem. The aim is to show something of the way in which work on the problem proceeded. How, for example, was the original question put into statistical terms? Why into those terms? How were standard techniques brought to bear, or new ones developed? How did the investigator's ideas change as the work progressed? To what extent did the statistical analysis throw light on the original problem? G. H. Hardy once advised authors of mathematical papers to beware of writing up their discoveries in the order in which they were made. Case histories – admittedly serving a different purpose – fly directly counter to his warning. By revealing more about motivation for the different steps in an investigation than is usual in more polished final accounts, and more about the way in which misconceptions and false starts were corrected, they provide the student, it is hoped, with a greater understanding of how an investigation is carried out, and with a pattern that he himself might follow in the future. Most of the statistical abilities listed in Chapter 2 could be illustrated through appropriately chosen case-histories, but the determination of the aims of an investigation (2.2.2b), the translation of general aims into specific problems (2.2.2c), and the interpretation and/or utilization of the results of the analysis (2.2.2h) are likely to figure prominently in almost any example. A course made up of several varied case histories might also illustrate the breadth of applications of statistical ideas and the variety of formal techniques that an investigator uses during his work, and it might convey this illustration much more rapidly than alternative methods of teaching which call upon the student himself to work through a similarly varied range of examples.

On the face of it, therefore, case histories offer some worthwhile benefits in the teaching of practical statistics. They do however suffer from a major, and perhaps overwhelming, limitation: they cannot exercise those skills of statistical practice which are active in nature. In a case-history course the student is cast in the role of a spectator, who bears no responsibility for formulating the problem or for deciding how it will be tackled, and who gets no opportunity to carry out the analysis – and to face up to the inevitable unforeseen snags that will arise during its progress – or to present a report on the work. Though a good teacher may be able to convey a convincing and valuable picture of the background to these activities, there is no substitute for direct experience. As a result, teachers may well feel that the case-history technique in the form described here is most useful as a preliminary to other, more demanding, courses which do exercise the active skills of statistical practice. Case histories may have a place, for example, at elementary levels, or when students lack confidence in applications, but are unlikely on their own to teach practical statistics adequately.

The difficulties of mounting a course of case histories centre, from the teacher's point of view, very largely on the problem of finding suitable material. For maximum benefit it seems essential to give some account of what went wrong in the investigation being described, and of what alternatives were considered, yet few published reports give this kind of detail. In general the teacher may have to fall back on his own experience or be faced with the laborious job of compiling histories from the oral accounts of other statisticians. Another difficulty is the problem of assessment. If the aim of case histories is to convey an appreciation of the way various experts tackled some particular statistical problems, then the form of assessment should naturally allow the student to demonstrate this appreciation. Questions eliciting verbal, as opposed to mathematical or numerical, replies seem most appropriate: 'What considerations led to the adoption of a correlated-errors model in such-and-such a problem, and how did this affect the analysis?'; 'Discuss the relative merits of a likelihood and a least-squares analysis for . . . '; 'Why were the conclusions from ——'s analysis of —— incomplete, and what further data would be needed to complete the investigation?' . . . Here, as elsewhere, a good understanding may be hard to distinguish from a good memory. If, on the other hand, the case histories are seen mainly as a prelude to some other course designed to exercise the active skills of practical statistics, then it might be reasonable not to attempt any direct assessment at all of the case histories but to take performance on the other course as an indirect reflection of the lessons that the case histories have helped to teach.

3.11 ROLE-PLAYING AND CASE STUDIES

The case histories of Section 3.10 suffer from the major drawback that they involve the student in statistical investigations only as a spectator, and not as a participant. A natural way round the difficulty is to assign the student a more active role as investigator for all or part of the problem. The result might be called a role-playing teaching technique, or, following Kanji (1983) and Bissell (1975), a case study – although the term 'case study' might also be used for what we call a case history.

A typical role-playing project or case study will be based on a case history. The teacher introduces the problem in a non-technical unformulated way and then the students, either singly or in small groups, act, in effect, as 'consultants', asking questions, formulating approaches, working on them, and finally presenting a report. The work may extend over several weeks. It may divide naturally into distinct phases, for any one of which the teacher might revert to a case-history approach of describing what was done in the model investigation, and why, with a resulting saving in time. Whether this stratagem is used or not, it will be useful to convene intermediate meetings to monitor progress and discuss the next phase of analysis. In all of these activities the students are meant to keep up their role as investigators, and a way of aiding this is for the teacher also to assume a role – at least for part of the time – as, say, the non-statistical poser of the original problem. If initial reflection leads the 'investigators' to seek clarification or further background information or more data, then they can return to the supposedly non-statistical 'problem-poser' with appropriate requests, and they can also be asked to explain their final conclusions to him in non-technical terms. With a little modest acting ability on both sides, it might then be argued, the case study can give experience in communicating with a non-statistical client as well as practice in many of the other activities of a statistical investigation. This perhaps overestimates the ease with which a statistician can adopt the necessary role: in practice it is surprisingly difficult to shed one's training convincingly. There is, moreover, another difficulty arising from the teacher's playing of the investigator's role, at least for more advanced problems: his limited knowledge of the background to the problem. In real practice a statistician often needs to ask questions which are not on the surface closely connected with the data: who draws up the nurses' duty roster and how; are the test pieces drawn from store in batches; is it possible to get further information – even at this stage – on the patient's history? The teacher is unlikely to know the answers to all possible questions of this type, and he will therefore be reduced on occasion to choosing either to construct an entirely arbitrary answer or to say that there is no information. Proposals for the improvement of future investigations will be particularly affected by this kind of ignorance of what is and what is not practicable. Thus it may be difficult in practice to carry off the role-playing part of a case study totally convincingly.

In any use of the case-study technique an important question that arises is: how close should the work keep to that in the original case history? Evidently to abandon the model investigation completely is to lose the possibility of learning from another's greater experience and judgement. On the other hand, if the teacher aims to constrain the 'investigators'' activities so that they mimic those in the case history, the exercise and development of the students' own skills is inhibited. There is an inescapable conflict here, and no solution appears fully satisfactory. Any attempt by the teacher to influence the course of the investigation – say by resuming his teacher's hat in place of his problem-poser's hat at an intermediate briefing session – runs the risk of underlining the artificiality of the exercise and of re-awakening the suspicion only lightly asleep in many a student's mind that the object of the exercise is merely to elicit from the teacher (whether wearing his disguise or not) the single key that leads to the 'right answer'. This, of course, largely defeats the purpose of the exercise, since what is learned then could be communicated more efficiently and with less apparent irrelevance by the case-history technique. A compromise is to have only one debriefing session at the end of the exercise, and to hope that some learning from another's experience will take place when the students hear – presumably with keener interest – about the approaches used in the real investigation. In this form, however, much of the affinity with a case study in the usual sense is lost, and the resulting teaching device is closer in spirit to the project-assignments to which the next chapter is devoted.

The extent to which a case study contributes to the development of the statistical abilities in Chapter 2 then depends very largely on its detailed form. It *can* give practice at working with others, working to a time-deadline, communicating clearly and efficiently (abilities 2.2.1a–c), and in principle it *can* also exercise abilities depending wholly or partly on technical skill, such as the ability to translate general aims into specific problems (2.2.2c) and the ability to apply any necessary techniques and to interpret results (2.2.3b, c). As we have seen above, however, it may not always be easy to ensure that these abilities are actually exercised effectively.

The problem of student assessment here is similar to that in any activity in which students collaborate, for at least part of the time, as for example in statistical experiments, discussed in Section 3.3.

3.12 SEMINAR PRESENTATIONS

The name describes the teaching method in which each member of the group investigates an individual topic or problem and subsequently presents an oral account of it to his fellows; in some cases he may also submit a written account. The method is common in some other subjects (geography for instance) and has been used as a final-year option in UK mathematics and statistics first degrees and also as a component of UK statistics Master's degree programmes. Topics might be techniques possibly useful to the applied statistician: transformations, probability papers, tests for normality, jacknifing, EDA techniques, for example; or, less commonly, investigations with a real data-set. We discuss the latter as an example of a mini-project in Chapter 4, and so concentrate here on the former.

The course may require each member to give several short – say from 15 to 30 minute – seminars during the year or term, or just one, presumably longer – up to an hour, say. In either case the benefit is mainly derived by the presenter, through his being obliged to read up a field on his own (though usually with considerable guidance from the teacher), order his thoughts on it, and present them to a non-expert audience. As every teacher knows, this is a good way to learn a new topic. There is also an undoubted increase in self-esteem and confidence associated with being 'the expert'. Evidently the statistical abilities from Chapter 2 most exercised by the activity are those of communication (2.2.1c) and those connected with the locating and critical reviewing of relevant literature (2.2.2g). If enough time for preparation can be allowed, it would seem better to aim at several brief seminars by each student, rather than one long one, since the student will learn as much about presentation from having to prepare a short talk as from having to prepare a long one, and by giving several talks will cover more subject-matter, as well as have the opportunity to learn from his mistakes. The benefit derived by the audience from the factual content of the talk is probably less in statistics than in many other subjects in which this teaching device is used. Our topics tend to be more technical and less easily intuitively appreciated, and that makes greater demands on both the presenter and the audience. The information gained by the presenter may therefore not be conveyed with full efficiency to the rest of the group, to whom the talks may seem little different, except for the distracting novelty of the speakers, from a rather disconnected lecture course. What members of the audience may learn – from the speakers' mistakes – are some of the essentials of good oral presentation: to speak audibly, to write legibly, not to obscure the overhead projector, and so on. If questions and discussion are the norm, members of the audience may also absorb, gradually over the weeks, an enquiring and constructive attitude to presented information. A mild encouragement for this would be a requirement that every member think up a question ready to put to the speaker at the end of his talk, a requirement enforced perhaps by an awareness that the teacher would choose a questioner at random when the time came.

In summary, therefore, the benefits overall from a course of this type are: experience at deriving, selecting, and ordering information from the literature; practice at oral presentation; a gain in knowledge of theory or techniques (mainly to the speaker); and possibly the promotion of an enquiring attitude in the audience.

We now briefly consider some of the organizational requirements of a seminar course. Firstly, how large a group is practicable? If the course is one in which individuals present a single talk at the end of the year, say, and otherwise have no contact with each other, and are under no obligation to attend each other's talks, then the only limits are imposed by the number of topics and the amount of time available for talks. If the group is to meet regularly, on the other hand, to hear each other speak, then, if it is too large, individuals will find they spend too much time listening and too little speaking to sustain interest and useful learning during the meetings. It may be advisable therefore to divide a large group into self-contained subsets. Exactly how large these should be will depend on how much preparation time is thought necessary for a talk and how many talks are scheduled each week: with two talks per week, for example, and a preparation time of four weeks, which represents a convenient work load, a group of 8 would be appropriate. Requirements from the teacher in a seminar course are, fairly obviously, suggestions of suitable topics, guidance on where to start reading and possibly on how to use bibliographic aids, and advice on how to prepare and deliver a statistical talk; useful references on the last are the papers by Freeman *et al.* (1983) and Mosteller (1980). During the seminar presentations themselves the teacher's role is likely to be quite modest, limited to putting speakers as much at ease as possible, and to encouraging discussion. Afterwards he may be able to offer individual advice on ways of improving presentation.

As with other courses in which students work at different tasks, assessment will be problematical. Topics will generally be quite variable in difficulty: for some there may exist connected expositions in the literature requiring little more than paraphrasing and condensing on the student's part; others may need a considerable effort of synthesis and ordering. In general it seems advisable to offer only topics which demand consultation of more than one source. Subjective grading of the oral presentations is possible, with an allowance for difficulty of topic. Greater objectivity might be achieved through the use of a panel of graders with an agreed breakdown of criteria of assessment. Even then, however, detailed discrimination cannot always be expected. On the other hand, assessment in this type of course is not needed as a spur to effort: the challenge of a public appearance is usually an adequate stimulus.

Cross and Moscardini (1985) have described a teaching technique which shares characteristics with both seminar presentations and case studies, and with the discussion groups of the next section. They used the technique in their teaching of general mathematical modelling, but a similar approach is in use in Germany for courses on practical statistics. In both versions the class of students is subdivided into groups of three or four, and a question is assigned to each group. At a subsequent meeting of the whole class each group reports back orally, via a nominated spokesman, on what members have found out about their question. Questions are generally on points of detail arising in some larger investigation assigned to the class as a whole. The class might have been asked, for example, to compare the utility of various goodness-of-fit tests and, having decided to use a Monte Carlo approach, might seek some detailed information about the method. One group might then be asked to find out about the generation of pseudo-random numbers from particular distributions, another about problems of cycle-length, another about variance-reduction techniques, and so on. The resulting presentations are often short and strictly limited in scope, since the questions giving rise to them are, and so do not make heavy demands on organizational or presentational skills. On the other hand, the audience is closely concerned with the subject-matter, and so discussion is likely to be generated naturally and spontaneously. Others of the statistical abilities in Chapter 2 might be exercised too, particularly through the discussion: the ability to work with others (2.2.1a), and possibly the ability to translate general aims into specific problems (2.2.2c), and to organize work (2.2.2e). A problem is that some individuals may leave the work within groups and in the general discussion to others. It is important too, as Cross and Moscardini emphasize, that the overall topic of the investigation be chosen within the capabilities of the class: too hard a topic can lead to discouragement and the cessation of learning.

3.13 DISCUSSION GROUPS

By this we mean the discussion of a practical question by a group of students, usually, but not necessarily, accompanied by a teacher. If the teacher is present he would probably not wish to dominate the proceedings but rather to act as a prompter and questioner, since the way that learning occurs in such groups is largely through the active involvement of the participants with the task in hand.

Group discussion may be used as a particular teaching device within some of the other approaches to practical statistics described in this chapter, particularly those based on critical reading, case studies, and seminar presentations (Sections 3.4, 3.11, 3.12). In this section, however, we concentrate on discussion groups in their simplest form, which we take to be the regular meeting of a group of people to discuss a range of statistical questions.

Not all types of statistical question, of course, are equally suitable for use in a discussion group. Problems in which the main points of interest arise only after fairly substantial numerical work, or those in which there is little room for discussion about the best method of attack are perhaps better regarded as material for a mini-project assignment (see Section 3.14) or a drill exercise, respectively. The planning of an experiment or survey, the best methods of analysis of fairly complex designed experiments or of substantial data sets arising from non-experimental work – such as the elections data discussed in Section 4.6.4 – all seem suitable, but these categories are by no means exhaustive.

The teacher's role in the group is, as has been suggested above, a rather delicate one. On the one hand he will hope to guide the discussion away from irrelevance and error, but on the other he will not wish to dominate the proceedings with his own ideas. In a sense it is not too important which technical points arise in the discussion and so he should not feel obliged to guide the deliberations towards some preconceived goal. What is more important is that a reasonable number of ideas should be put forward and that they should be subjected to rational evaluation by the group. The habit of looking for and examining the advantages and limitations of any idea is perhaps the single most important skill that can be learned in a discussion group. This being so, it is not of overriding importance that the group should reach a single definite conclusion about the problem it has been considering, though its proceedings will be better focused if members regard that as the intention, and there will probably be a greater sense of satisfaction if the meeting ends with the drawing up of some fairly specific recommendation.

The following is typical of the way in which a group could operate. The teacher will put forward a question, often accompanied by some data, and ask the group to draw up a scheme for its analysis. It would usually be most efficient if the problem were given out some time before the meeting of the group, so that members could give some thought to it in advance. It might be useful, too, for the teacher to hint that every member will be expected to have something to say at the group meeting later, thereby encouraging those who might otherwise hope to take a back seat to find time for some preliminary thought. The way in which the meeting proceeds will depend on various factors: how large the group is, how relaxed in each other's company the members are, how familiar they are with similar kinds of meeting, how much they have thought of in advance to say about the problem . . . . In some cases, especially when members are used to this kind of activity, the discussion might proceed reasonably fruitfully with little intervention from the teacher. However, more usually, and especially in the first few meetings, the teacher will probably have to guide the proceedings, taking the role of a rather retiring committee chairman: posing individual questions for consideration and summarizing conclusions reached, but otherwise leaving members to make their own points with little interference. If he wishes to reduce even this level of apparent involvement he might try adding some intermediate leading questions in advance when posing the original problem, and nominating other members of the group in turn to be responsible for starting the discussion at successive meetings.

There are various practical points about arrangements for meetings of a discussion group that it might be useful to note. One is that we have found that arranging the seating into a rough circle, so that members can see each other easily and so that the teacher does not occupy a distinguished position, is beneficial. It is useful, too, to have a blackboard or overhead projector available so that formulae or graphical arguments can be easily displayed. On occasions we have found it helpful to suggest that a member of the group should be elected to serve as a secretary, with the job of recording conclusions agreed upon. This has the effect of giving a gentle and continuing reminder of the supposed aim of the meeting and of discouraging irrelevance, without the teacher's direct intervention. An ideal size for a group is, in our experience, in the range 6–12: with more than twelve there is a tendency for some members not to become sufficiently involved.

The assessment of an individual's performance in a discussion group is difficult. One view is that assessment should not be attempted at all, since it will interfere with the primary purpose of the group, which is the development of certain skills by participants. However, if some assessment of individuals is required it might be obtained by asking each participant to write a report listing and evaluating the points made at the meeting. The report would not provide much evidence of the individual's originality, since the points would not necessarily be his own, but it would allow some assessment of his ability to appreciate and evaluate rational arguments. A more direct form of assessment that has been used in some discussion groups is for one or two experienced observers to listen to the discussion and grade each participant's contributions for originality, relevance, tact, etc. There is a very strong danger with this method, however, that discussion will be inhibited.

What then, are the advantages and limitations of group discussions for the teaching of statistical practice? Suppose that a course on practical statistics includes a substantial number of discussion group meetings on a variety of problems, say one problem per week for a term or more, with the possibility that occasional problems may require more than one meeting. By the end of the course members of the group will have met a variety of data sets and examined, in general terms at least, strategies for analysing them. If the projects were well chosen, members will have built up a valuable fund of technical experience, on the level, say, of knowing what considerations are involved in designing particular classes of experiments, how a least squares estimation problem in which parameters satisfy inequality constraints might be tackled, and so on. At a deeper level, too, it might be hoped that they would be more ready to put forward and examine objectively ideas for approaching practical problems, and to work in a cooperative and constructive spirit with the rest of the group. On the other hand, there is a price to pay for the relatively large variety of problems that can be considered by this means: for none of the problems is the actual analysis carried out. This is quite a serious limitation, since for many real problems much of the challenge to understanding and interpretation becomes fully apparent only during the course of the analysis itself. The student trained in practical statistics only through the present approach is in danger of becoming an experienced armchair strategist who has never fought a battle.

The extent, therefore, to which the pure use of discussion groups provides for the requirements of a statistician as listed in Chapter 2 might be summarized as follows. It may give direct practice in the choice and development of an appropriate plan or design for an investigation (2.2.4c) and in the initial stages of building a model (2.2.4f) (but not the later, refining, stages since the analysis is not carried through) as well as in the recognizing of which techniques are valid and/or appropriate (2.2.3a). Others of the qualities in Sections 2.2.2–2.2.4 may be engaged and developed more peripherally, depending on the particular problems discussed. Of the more generalized skills listed in Section 2.2.1, the discussion group will give practice in that of clear communication, orally at least (2.2.1c), and in those skills contributing to successful cooperation with others (2.2.1a). With appropriate constraints on the work of the group it might also give experience of working to a deadline (2.2.1b), and in recognizing that a less than perfect answer might have to be accepted as a consequence (2.2.1d). What the use of the discussion group will fairly emphatically not do is to allow practice of those very important skills needed for the numerical part of the analysis of a problem and in the interpretation of results (2.2.2h and 2.2.3b, c).

3.14 POSTSCRIPT

Though the foregoing sections describe a wide variety of ways of teaching statistical practice, they are by no means exhaustive. Activities which, for instance, combine features from two or more of the approaches can easily be found. The German version of a seminar presentation course described in Section 3.12, for example, might be regarded as a kind of hybrid with discussion-group activities, and also has similarities with the case studies of Section 3.11. Role-playing, discussed in a rather specific sense in Section 3.11, enters in a general way into several of the other methods – indeed whenever the student is asked to imagine himself in the position of a working statistician – and, given the aim of encouraging both specific abilities important to a working statistician and imaginative qualities more generally, it is natural and desirable that it should.

By selecting particular methods or devising activities based on a mixture of methods, the teacher can aim to produce courses tailored to the needs of different groups of students. At an elementary level, for instance, some types of activity demanding a relatively broad technical background (case studies and discussion groups, for example) are likely to be relatively unprofitable, and others which lay the foundations for more complex activities later (drill exercises, simple seminar presentations, case histories, for example) are probably more useful. At a more advanced level, when much of the standard *technical* equipment of the working statistician can be assumed, work which more directly contributes to development of the higher-level skills of Chapter 2 should be possible. Much of the groundwork behind these skills – in the form of a modest ability to communicate, a general scientific awareness, some flexibility of mind – will have been laid earlier, and the need at this stage is to develop these same skills further and to encourage their use in combination; thus it seems highly desirable that the student should be confronted now with real problems which simulate as closely as possible those facing practising statisticians. For most efficient learning it seems essential that his involvement in these problems should be active as opposed to passive – that he should take the ultimate responsibility for deciding on an analysis, and that he should carry it out and follow it through to the stage of reporting on results. Also, since the problems faced by practising statisticians are varied, and different problems demand the exercise of different skills, it seems highly desirable that the student should work on as wide a variety of such problems as possible.

These considerations suggest that a course based on a varied sequence of realistic short assignments to be completed by the student is likely to be particularly effective as a culmination to a practical statistical education. It is likely too to be valuable for students nearing the end of their training who have had little other explicit teaching in practical statistics, since it lies closest to active work experience. We will call such a course a 'short-assignments' or a 'mini-projects' course. Its underlying technique is the old and simple one of giving practice in real problems, with guidance as they are tackled and constructive criticism afterwards. The detailed teaching methods used in particular assignments may be varied to suit the problem and to emphasize different skills: some assignments may be conducted partly as case studies, others by means of group discussions, seminar presentations, critical reading, and so on. The differing advantages of the various methods may in this way be exploited, and variety and interest maintained.

This short-assignment framework for a practical statistics course is therefore to be seen not as an alternative or addition to the teaching methods discussed earlier in this chapter, but as a simple and effective way of combining and developing some of them for particular needs. The student is able to learn in this framework both through the experience he gains in working on each assignment, and also through the feedback provided by the teacher's comments; the sequential nature of the activity also provides for consolidation and reinforcement of what has been learnt. This framework is highly attractive from an educational point of view and, moreover, is flexible enough to be used at less advanced levels too: at any level, all that is required is that the work in individual assignments be adapted to the students' capabilities. A mini-project course therefore gives a simple, effective and flexible way of building on the strengths of the various teaching methods reviewed earlier in this chapter. In the next chapter we discuss its use in more detail and in the following chapter give examples of assignments suitable for implementing it at various levels.

CHAPTER 4

Project-based courses

4.1 INTRODUCTION

What we were doing in Chapter 2 was a broad version of task analysis, for practising statisticians. The fact that the statistician we are hoping to train may go on to work in a variety of areas – in government, in medicine, in agricultural research, in business, and so on – makes it impossible to make the goals much more precise, except that one could go much further and list the technical material which more or less all statisticians should have mastered. Our particular concern here, however, is in ensuring that the wider issues are properly dealt with, and in this chapter we shall discuss how the teaching framework can be used to foster the broader abilities desired.

In Chapter 3 we have discussed a variety of approaches to teaching the practical aspects of statistics, indicating where the strengths and weaknesses lie. In the present chapter we discuss a type of course – a method of teaching if preferred – which combines and extends several of these, going beyond any one of them, though building on their strengths, and thereby providing a more powerful way of teaching statistical practice. The fundamental characteristic of such a course is that from the beginning the emphasis is on tasks which are at least approximately realistic: thus the course is made up of a sequence of *short assignments* each of which is:

- posed in the form of a problem in the real world, rather than in statistical language.

As a contribution to the reality there is an insistence that each assignment be

- completed to a strict time deadline; and
- reported on in writing and perhaps orally, in subject-matter rather than statistical terms.

The assignments are carefully chosen to be realistic, open-ended and not technique-oriented. (They must, of course, be appropriate to the students' statistical and general maturity.) They often do not give all the information needed, and sometimes some of the information that they do give is not relevant. Students are encouraged to interpret them imaginatively, seeking further

52

information for themselves if necessary and recasting the questions into alternative forms if they judge that to be helpful. Work on many of the assignments will require application of standard techniques such as regression, analysis of variance, etc., as well as use of the computer, but the purpose of the course is emphatically not to give drill in these: the assignment questions are so chosen that, even when the report finally submitted is very largely based on, say, a GLIM or MINITAB regression analysis, some serious thought and judgement will be needed (and will have to be fully discussed in the written report) before that particular technique is chosen as an appropriate tool for that particular question, and the way in which the formal output from the technique is used to answer the question will also require careful consideration.

One point should be emphasized at this stage: we regard such a course as losing much of its value – indeed its justification for being called a *course* – if the teaching and learning aspects are ignored. The reports on the assignments are not, therefore, merely to be handed in and returned, if at all, very much later: they are to be marked (annotated), in the sense of constructively criticized, in order to provide feedback to the student on how he might do better in his next task. We shall comment later on marking for assessment or grading purposes.

There is no reason why virtually all of the teaching approaches discussed in Chapter 3 should not be used within such a framework. In particular, if one has the time and resources, there is a great deal to be said for completing a programme of the kind we advocate with a final large project, individual to each student (at least if the project is carefully chosen), for only with a large project can one hope to give a student practice in the continual refinement of the approach, as preliminary analyses are completed and as the difficulties show themselves, that would be necessary in much real statistical practice. At the other extreme, drill exercises of various kinds are essential as new ideas are presented; but it is also essential that these do not give the impression of being ends in themselves. Case histories, again, can be very valuable in showing how other people solved certain problems, and in opening up, for some if not all students, a whole landscape of new possibilities.

4.2 THE NEED FOR PROJECT WORK

It will be convenient to refer to these tasks which simulate reality – to a greater or lesser extent – as projects, though in view of the use of the word in connection with what we called 'large projects' in Chapter 3, more appropriate if also more clumsy names for most of them might be 'mini-projects' or 'short assignments'. Examples of such mini-projects are given in Chapter 5, and a number are discussed at length in Section 4.6, but for convenience we reproduce here, as Exhibit 2, Project 8 from Chapter 5.

Projects – large projects, that is – have frequently been discussed as a means of providing valuable experience, not obtainable in other ways, in a wide variety of

EXHIBIT 2

DELINQUENCY AND FAMILY SIZE

The following two letters appeared in successive issues of *The Times* in January 1973, and are reproduced by kind permission of their authors. Names have been changed.

From Mrs Beta: Sir, I was interested in the final point in Mr Alpha's letter entitled 'Unwanted children', where he wondered whether delinquent children more often came from large or small families. I sit as a magistrate on the Inner London Juvenile Court Panel and this same question occurred to me last summer.

Accordingly during my last quarter's sitting I have kept a tally of the number of children in the families of children in court whenever this information was given to the court. It is as follows:

Children in family	No. of cases
1	3
2	3
3	9
4	16
5	8
6	15
More than 6	16

I wonder whether juvenile court magistrates in other parts of the country have noticed a similar trend.

From Mr Gamma: Sir, your correspondent, Mrs Beta, who infers that delinquent children more often come from large families, would appear to have misread her own information. Does Mrs Beta not realize that, quite apart from any other factors, the average family of six children must stand six times more risk of raising a delinquent than a family with only one child? If the information that is given is recalculated to eliminate this loading, the table will appear as follows:

Children in family	No. of cases
1	3
2	1.5
3	3
4	4
5	1.6
6	2.5
More than 6	Less than 2.3

Draft a letter to *The Times* commenting on the statistical logic of Mrs Beta and Mr Gamma. You may wish to refer to information contained in Table 2.8 of *Social Trends* 1975.

Since such a letter must be relatively brief, set out your argument more fully in a longer report, incorporating any further analysis that you think appropriate.

subjects. It appears, for example, that, in engineering and architecture – and very likely other – undergraduate degree programmes, project work is almost a universal requirement, which may perhaps be due to the fact that these subjects are always seen as intrinsically applied rather than academic in nature. Large projects are very common in postgraduate taught courses (master's programmes) in statistics, but, while not unknown, are not especially common for under-graduates, and are probably unknown in service courses for other departments. (A recent discussion of the use of projects in teaching, relating mainly to electronic and electrical engineering but containing a great deal of information and giving many references, has been given by Allison and Benson (1983); Griffiths and Evans (1976) and Kanji (1979) have some interesting things to say in the statistical context, and a recent conference addressed the subject (Croasdale, 1985a), but these last are perhaps the only references on the subject written specifically with statistics in mind.)

The idea of smaller projects seems much less well known. Clements and Clements (1978), discussing engineering mathematics, observe that service courses are typically bedevilled by the way that students compartmentalize their studies, so that, even if real illustrations and applications of mathematics are described, there is little or no synthesis of the two subjects; that is, the students cannot readily be persuaded to think mathematically outside a narrow mathema-tical context. Their solution is to recommend what they call simulations [based on] case studies – which happen to be of exactly the same type as some of what we call projects. Chatfield (1982) is an isolated reference to the use of small projects in statistics.

The virtues of projects stem largely from the involvement of the student in some activity which could have, and perhaps even does have, some real and lasting end-product, and the more this is the case, the greater the educational value is likely to be. 'Learning by doing' is always acknowledged to be better than passive absorption of knowledge, and in this situation the learning that we are seeking is, in the end, how to be a practising statistician; and so the nearer the project to real life, the better from this point of view. But in addition to efficiency in this sense, most students find it more interesting to do this kind of work than to spend all their time re-proving results that are already known, or calculating answers which have no intrinsic interest. (Not all students think this way, of course, and those who are irredeemable purists may well regret the time spent on doing projects, particularly the writing of reports.) To see the relevance to the real world of material covered in lectures is not only neutrally interesting, moreover; it helps to build confidence in the ability to apply the theory and methods already studied, and it has incidentally the effect of assisting the student's personal development. We shall go into much greater detail later on some of these aspects, but we note here that projects require decisions for which responsibility has to be taken: which method, how much effort, and occasionally a more obviously irreversible decision, as in experiments or surveys; cooperation (in group projects); organization of resources (always time; sometimes others); communication both oral and written; extraction of relevant information; and understanding of subject matter in other fields, to name only the obvious ones. It will be noted that these are connected with the desirable abilities listed in Chapter 2, and also that they are not fostered to any significant extent by lecture classes and most of the other traditional types of teaching method. The qualities needed in projects are in fact different from those contributing to more conventional work, particularly those assessed in short unseen examinations, in which someone who thinks deeply but slowly is at a considerable disadvantage, and inventiveness and originality can be demonstrated only in a rather limited fashion. It may even be hoped that by developing these qualities such tasks strengthen the ability of a student to teach himself, and thereby the ability to keep learning, by himself, after the end of formal education.

It might be argued, as we commented earlier, that it is unnecessary to pay any attention to more general qualities of this kind in, say, a university course: but in fact it is just such items that are generally reckoned to make the difference between training, in a relatively routine sense, and education. Sims (1976) noted, in rather different words, that one of the characteristics of the professional engineer is that he will use broader and different skills, indeed different kinds of skill, as his career progresses, for otherwise he is really only a skilled technician, and that it is worth getting some idea of these more general matters at an early stage; and every word of this surely applies equally to statisticians.

At least in principle, then, project work has great appeal, in encouraging a much wider range of personal skills, in creating interest, and generally in showing students what the practice of statistics is really like, though of course to achieve these ends without sacrificing something from the technical side of statistical theory and method is not easy, and may be impossible. The main aim of the rest of this book is to show how, in detail, such project work can be introduced into statistics programmes.

Before we go further, however, we may as well face the fact that there are drawbacks, as well as advantages, associated with its introduction, for both teachers and students. For teachers, or at least those trained in the traditional way, it is harder, and it is certainly more time-consuming, to mark reports in a helpful way than to mark relatively routine exercises. Students also find project work hard, particularly if they are, as is common in many countries, mostly mathematicians in origin, since they are then probably not in the least accustomed to writing in a natural language; in any case the translation of real problems into statistical language is a hard task. It is also very easy for the teacher to impose too heavy a work-load unthinkingly. Not only are these points important in themselves, but if statistics is a subject which can be chosen or not, there is a danger that students may avoid statistics because of them. The advantages are so great, however, that, although of course one must keep a careful watch on the situation, one should be able to persuade the students that the approach is both more interesting and more beneficial – for example, in preparing them for future employment; indeed one can easily and truthfully argue that they will find the experience useful no matter what their eventual career, and it therefore seems to us that it is right to argue that projects should be compulsory for those who do choose statistics, rather than optional. A few students go to the other extreme, and become so enthusiastic about this kind of work that they willingly neglect the rest of their studies, which is just as undesirable.

4.3 PRACTICAL ARRANGEMENTS

We shall give examples of projects in Section 4.6 and in Chapter 5, but for the moment we restrict ourselves to the mechanics of how to use them, and in the next section we make some suggestions about improving students' general abilities. At various points, incidentally, we make rather specific recommendations (about the number of projects, for example): these are based on our experience, but we would not be too surprised to hear that others find rather different arrangements preferable.

Before going into details, we repeat that we see the teaching and learning aspects of such a course as paramount, and so the practical aspects should be determined to support them. Students learn in two ways, at least:

- through doing the projects – working out what is needed, carrying out analyses, writing up the report, etc.;
- through constructive criticism of their reports or indeed of their ideas.

Recognition of this dual mode of learning helps the teacher decide how best to organize the course. It follows, for example, that teachers should be prepared to offer advice on particular projects when asked – though not, of course, advice of the detailed prescriptive kind.

For convenience we shall discuss practical arrangements in the context of a traditional three-year English university degree in which at least half of the final year is devoted to statistics, but it is not difficult to adapt the general considerations to other levels or lengths of course; naturally the projects themselves need to be chosen appropriately – indeed, even within a given framework it is important that the projects should be chosen to match the students' technical knowledge and maturity, so that what one is looking for is a sequence carefully chosen to display increasing and appropriate difficulty and challenge. Even for service courses – in which statistics is taught directly as a tool to be used in the service of some other discipline – it is worth trying to include some project work, though this may not be possible with really short courses.

The successful introduction of a course including substantial project work necessitates considerable advance planning, with decisions to be taken about all kinds of matters. In the first place, while the students have relatively little knowledge of statistical technique and theory – for example in the first year and perhaps in the second also – the projects may as well be attached directly to a course on methods, and probably not too many, perhaps four to six a year, should be set. However, in the later stages a separate 'course' consisting entirely of projects is desirable (and it should then be a co-requisite with methods courses): firstly because it is right to expect rather a lot of this kind of work, and secondly because it is crucially important to avoid by this stage any suspicion that the intention of a project is to illustrate a particular technique. What proportion of the whole year's work such a course might represent will inevitably depend on a number of local considerations. But for the final year of a degree course it might well be in the range one-quarter to one-sixth: this would certainly mean that some interesting and useful methods, quite apart from theory, cannot be treated because of limited time, but this will be true anyway. Anything much less than this will unreasonably limit the range of practical work that a student undertakes, however.

In order to provide variety in the type of project, a reasonably large number of projects should be set: depending to some extent on the length of the teaching year, something in the range 10 to 15 seems sensible. Unless it is made a requirement that for an overall pass a satisfactory attempt must be made at each project, there is inevitably some choice for a student: he may omit a project and thereby accept a lower potential maximum mark. But we suggest that the understanding should be that normally all projects will be attempted, so that it is not possible to avoid the variety of types of project.

In the preceding paragraph we wrote as though it were inevitable that each project would be assessed as well as annotated, and that the assessment would then contribute to the final grade: but in fact whether this should be so requires a difficult decision. In favour of it are the facts that to examine the material covered is impossible in a small number of conventional examinations, and that as the projects are quite hard work they will get less attention than they deserve if they are not assessed – and in any case a student who works hard at such a course ought to get some credit for it; against are the two well-known drawbacks of continuous assessment, that it confuses learning with assessment, and it puts permanent pressure on students. We ourselves feel forced into making the individual assessments contribute to the final grade – but wish we could find a satisfactory alternative. A slight variant, provided institutional rules allow, is to use project work as a test for permission to enter other examinations, rather than to provide an actual mark. It is of course possible to have an examination as well as graded projects: such an examination might consist of problems which are rather similar to, if shorter than, the projects, or it might have rather different aims. (Some institutions have, incidentally, customarily had very long practical examinations – 7 hours, for example – with, naturally, a supervised break in the middle.) We return to the assessment of projects later in this section.

It is scarcely possible, and probably not desirable, to imagine projects of the type we have in mind being done *in class*, so it is important to decide how they shall be organized for the greatest benefit of the students. A clearly specified date and time for the final report to be handed in is necessary, and any lateness without good cause should be severely penalized: in some circumstances perhaps it may be appropriate to refuse to accept late submissions, but at the very least a substantial reduction – by 25 to 40 per cent per day or part of a day late – in the mark otherwise achieved should be made. The teacher does of course have a corresponding obligation to mark (both to grade and to annotate) and return the reports as soon as possible, on obvious grounds as well as on the slightly less obvious one that if too long a gap ensues the students, unsurprisingly, will have largely lost interest in the details, and an opportunity to learn from mistakes will have been lost.

Exactly what else one should insist on doing depends substantially on the type of project, on the number of students in the class, and on the time allowed for each project, which in turn depends on the number of projects to be done during the year. (It would in principle be possible to give out ten projects at the rate of one a week, and allow ten weeks for each, but unless one has quite remarkably well organized students this will lead to chaos, and it is of doubtful value in any case.) With projects which are to be completed (at least) two weeks after being given out, it is useful to meet after one week to allow the students to put forward their ideas so far as they have yet evolved. The appropriate level of input by the teacher is difficult to judge, but a socratic style is very likely best: taking the students' ideas and encouraging the class to develop them, in the meeting, one stage further to see if they still make sense, or even just asking why it is thought that such and such an assumption is appropriate. Certainly a formal lecture is very unlikely to be the best format. Again, to have students report orally to the class as they hand in their project reports is excellent training for them. Both of these activities need careful organization by the teacher, unless the class is very small indeed, since students are commonly reluctant to say much, presumably for fear of being wrong. It is convenient for one or two students to be nominated in advance to give such short reports. If by the time the course is finished all students have spoken, so much the better; if this is not possible, then a choice will have to be made, perhaps at random. If possible, incidentally, other staff should be encouraged to come to listen, as should certain other groups of students – probably including those who are likely to be doing the same course in the next session; this may not be possible throughout the session, but if, for example, the plan mentioned below is adopted, in which the final project of the year is on a rather larger scale, then it may well be convenient to arrange a seminar afternoon (or day, or days, depending on numbers) for the oral reports on these longer projects, at which wider attendance is encouraged. Whatever the details, one should encourage the audience to ask questions: the ability to react quickly and intelligently to questions is undoubtedly both useful and improved by practice – as indeed is the ability to *ask* questions.

In order to provide a variety of experience it will be necessary to set a variety of project *types* (this is discussed in greater detail in Section 4.5); unless, however, the class is very small, for a given project type one has little choice but to set the same project for each student. Though this can ease the labour of entering data when necessary into the computer, which can then be a shared activity, it does raise the question as to how much students may collaborate with one another on subsequent work on the project. Too much discussion might mean that some students do very little thinking, preferring to copy others' ideas; too little would be regrettable in that students can and should learn from each other. If the assessments are to count towards the final grade the possibility of a student getting assistance is quite likely to give the teacher concern, at least initially, but our own conclusion is that it is not a serious problem. We explain to the class at the beginning that discussion is a good thing, but that the final report should be the individual student's own work, and point out that it is to his own advantage not to rely too much on others. Provided that direct copying of reports does not take place, which is easily checked, and provided that success in the project does not depend largely on a single flash of inspiration, which does not in any case seem very appropriate, the amount of thought needed, even with collaboration, will probably provide adequate benefit for the student, and also allow sensible individual assessment. If projects are to be done outside class, there is in any case no way of preventing a student from consulting an expert, or, if projects stay the same from year to year, a student from a previous year.

Our own practice has been to have all projects common except the last, in which students, either singly or in pairs, all tackle different questions, and, as these are also somewhat more substantial, they receive double weight in the overall assessment: the prospect of this also encourages students to treat the earlier projects seriously, as practice.

It is, incidentally, important to think out carefully how the projects are to be introduced to the students. It seems appropriate to give a short introduction at the beginning of the course to the purpose and nature of projects in general. This could include, for example, a brief reference to the various types of project, to the kinds of skill which projects are intended to develop, and to the way in which they are meant to be tackled – that is, imaginatively, with the student encouraged to seek further information for himself if necessary and even to recast the detailed question into an alternative form, with due justification, if he judges that to be helpful. (Though this last is not to be taken as a licence to ignore the problem as originally posed, and a full explanation would be essential.) A wise piece of advice from Chatfield (1982) could also bear emphasis early in the course:

> Instead of asking 'What technique shall I use here?' . . . ask
> 'How can I summarize these data and understand them?'

In the same vein, the student should be encouraged always to ask himself what kinds of results he expects to get out of any analysis – must all regression coefficients turn out positive, for example, for a model to make physical sense (as they must in the project discussed later in detail in Section 4.6.4); do general considerations suggest that a fitted curve should have a minimum in a certain region; and so on.

Subsequently in the course each individual project assignment will be based on a written project-sheet of the kind illustrated in Section 5.2 (the 'comments' there are not on students' copies), supplemented by an oral introduction. The latter will depend on the project, but will typically discuss further why the problem posed is of interest, and paraphrase it; it may discuss other aspects which seemed too general to put on paper; and it may contain further information relevant to the project – for example, in connection with Project 23 of Section 5.2 it may be sensible to talk briefly about *Nature*, and Project 7 provides a convenient place to indicate the range of government publications and to mention bibliographic indexes and other sources of further information. On the other hand it is essential, in those projects which may involve the choice and use of particular statistical methods, to avoid, at this stage anyway, remarks, perhaps in answer to questions, which point in particular directions: we know from experience, for example, that it is very easy, in connection with Project 27, on French election data, to find oneself hinting at regression. Some advice and assistance may turn out to be necessary at some stage, and it is a difficult course to steer between giving too much and too little. The guiding principle in this connection is that comments from the teacher should aim to prompt thought and to stimulate the exercise of the skills which the course is trying to foster, rather than to provide ready-made answers which lead only to routine activities.

Finally, in this section, we return to the question of assessment; in particular what should be assessed, and how the assessment should be made. It is widely agreed that, in so far as it is practicable, more than one aspect should be assessed, and among those put forward as providing potential contributions to the final grade have been enthusiasm, commitment, imagination, progress reports at intermediate stages, participation (in group discussions, for example), oral skills of presentation, and of course the final report. On the whole it seems to us that, to the accuracy that one can measure such things, the first four will reappear in the others, at least for shorter projects, so that one might well restrict oneself to the final three. How far one can, or should, assess participation and oral skills depends on the balance between amount of project work and size of class: in a large class participation will be somewhat difficult for the students, and also difficult to assess, since not all members of the class may have the chance to speak, and in such a case one is thrown back on the report. In any case the proportion of the final grade allotted to these other aspects will normally be fairly small – we would say no more than 15% for oral skills and 10% for each of the rest, and no more than 25% for these aspects in total – so that a rather crude mark scale will be both convenient and adequate.

As far as the marking of project reports is concerned, the situation we envisage, with any one student submitting several, or perhaps many, reports during his programme, requires slightly different consideration from that in which a once-only final report describing a 'large' project is submitted. In addition to providing information for grading purposes, it is most important, as we emphasized earlier, that marking also includes annotation in a form useful for students, not just because of the general desirability of openness and so on, but also, and crucially, because this is a way of allowing students to learn from their mistakes: thus technical comments, comments on the arrangement of the report, and criticisms of grammar, spelling, and English style are all appropriate. Some, for example Kanji (1979) and Croasdale (1985b), prefer to draw up an elaborate marking scheme; our preference is not to go to quite these lengths, but rather to use a fairly detailed check-list of things to be considered in arriving at a grade, in order to improve the consistency of marking. What is clear is that grading or annotation which concerns itself with technical statistical issues only is losing much of the value of projects, and is indeed missing the point.

Project work means hard work for the teachers as well as for the students, and for the shorter type of project that mainly concerns us here most of the teaching work is associated with the annotation and assessment; to put it somewhat differently, one needs to think rather carefully about manpower requirements, since project-based courses are much more demanding than traditional lecture classes. To mark project reports carefully so as to be of the greatest benefit to the students is time-consuming, and with anything other than a very small class there is a need to plan it carefully. Moreover, given the subjective nature of the assessments it is desirable, if it is at all practicable, to have two (or more) staff reading each report; with a larger class this may well mean several staff members each reading only a subset. It is of course useful to involve other staff in the assessment of the oral presentations if they attend them.

4.3.1 Problems associated with large classes

Although the benefits of project work remain much the same no matter what the level of student or the size of class, the resources needed (for reading written reports, or for listening to oral ones) vary considerably, and, in particular, to implement some of the suggestions above for large classes will prove difficult in practice. We have made a few remarks in passing on this in other sections, but consider the problem rather more systematically here.

It is not, in most respects at least, the size of the class as such that matters, since a class can always be divided into smaller sections, but rather the ratio of the number of students to the number of staff giving the course; sheer numbers might cause a problem in a few special cases – for example, more space might be needed than is available to lay out an experiment, or access to computers or libraries might be difficult for very large numbers.

What then are the particular problems that might arise with large classes, and how might these be alleviated? There are two components – the oral presentations and the written report – that are envisaged, and both will cause problems. The oral account is, presumably, always likely to be rated as appreciably less important than the written report, and in consequence one can, perhaps, largely ignore problems in this connection; as we suggested earlier, one could for example, decide that only a proportion of students can be expected to speak during the course, perhaps selecting at random those that are to speak. But one cannot justify treating the written report in the same way: it would be unreasonable to ask students to do substantial amounts of work and to write reports on this work if some or all of it were to be unread – and, crucially, not commented on – because only a proportion of reports, or perhaps of sections within reports, were to be read. (And it would certainly be unethical to grade students on partly or wholly unread material.)

In some circumstances the problem can be largely solved by using graduate assistants, but this solution will only rarely be possible. If assistance of this kind is available it will be desirable, if not quite essential, to set up a training/monitoring scheme to ensure that all assistants are looking for the same points and are giving adequate feedback. One might pursue this line of thinking to the extent of wondering whether self-assessment, or assessment by other students in the same class, is practicable. It would not be difficult to imagine a system of this kind in which rather specific aspects of the report were checked – is there a title? is there a clear, single-sentence, description of the problem? etc. – but much of the value would be lost by this, and, in any case, it is unlikely that students would think this an adequate response to their hard work.

Another approach to reducing the load to manageable proportions is to limit the amount of work to be done, by ensuring that project reports are not unreasonably long. It makes sense, in most cases, to set an upper limit to the length of a report (even for small classes), since in the real world no one will wish to read a report which goes on and on, no matter how remarkable the contents, and so for large classes one might well, with appropriately chosen projects, impose *severe* restrictions on length. The other possibility of this kind is to simplify the projects; since on the whole large classes are associated with the more elementary levels this may be a particularly sensible move. The simplification might consist of choosing problems which are in overall terms less complex, or it might be achieved by requiring the students to work in groups (see, for example, Section 3.12) asking for only part of the report to be written: as an example of the last approach the assignment shown in Exhibit 3 (Project 14 from Chapter 5) gives the introduction and the analysis sections of a report and invites the student to supply the discussion and conclusion; one could equally give a description and the analysis and demand the summary and introduction.

Quite apart from the virtues of reducing the amount of material to be read, incidentally, such a project has an additional benefit to recommend it in connection with beginning students: it reduces the single overall task to a collection of smaller tasks, on each of which the teacher and student can

EXHIBIT 3

STARTING A NEW JOB

You have just started work today in a new job as a statistician with an emerging young consulting firm. On your desk you find a small file of papers left by your predecessor on the day before his hurried departure to take up a more lucrative appointment with a rival firm. Among these papers are parts of an uncompleted report which your new boss – in the hurried five-minute conversation you were able to snatch with him before he left on a 3-week business tour of the Middle East – has asked you to finish. 'It shouldn't take more than an hour or two', he had said. 'All the analysis has been done – it's just a matter of tidying the thing up, putting in some conclusions, and sending a completed draft report off to the client, who by the way is clamouring for results. Don't waste time on any more analysis, but if any ideas strike you about what's already been done, or about the way it's been written up, jot down a memo about it for the file – in case we get any come-back. By the way, if you can get this off tomorrow, there's a good chance you'll be asked to present the results at a meeting in Paris the week after next.'

The partially completed report [which is reproduced in Section 5.2, Project 14] contains, you discover, an introduction, the data, and details of the analysis. Draft out further sections as your boss asked, and a separate memo containing any further comments for future use.

---

concentrate to the exclusion of other aspects. In some ways such projects are not dissimilar to *critical reviews*, as described in Sections 4.5 and 4.6.3. One can take this line of thought much further to give practice in very specific aspects: for example, at some (early) stage students have to stop saying that 'the result is significant at 5%' and start saying that 'there is some evidence against [the null hypothesis]', and this could itself provide the basis for tasks which are hardly projects, but do have some features in common with them.

4.4 SOME PARTICULAR CONSIDERATIONS

Among the skills that appear, explicitly or implicitly, in Chapter 2 and that are required in doing projects are three that are common to most types of project and that are sufficiently narrowly defined as to allow useful remarks to be made away from a specific context: the management of the student's time, the writing of a report, and the presentation of material orally.

4.4.1 The management of a student's time

In the last analysis only the student can arrange things so as to use his time efficiently and profitably, and of course some students are quite capable of organizing themselves without assistance, but observation suggests that many have difficulties; whether or not words of wisdom from the teacher achieve much is not obvious, but they can hardly do any harm.

The most obvious thing to say is that time is limited, both because the finished report has to be handed in after a week, or two weeks, or whatever it may be, and because, in order not to take time from other parts of a student's programme, some nominal time is allotted to each project. What a student should do, therefore, is to study the project as soon as possible, perhaps for some standard fixed period, so as to discover what he expects will need to be done: a visit to the library may be necessary, for example, the data may have to be entered into the computer, a first analysis – graphical or otherwise – may be called for, followed by further analysis, and so on. Since there are different kinds of activity involved this will give a first guide to starting a timetable: for instance, a visit to the library may mean going some distance across the campus, which can perhaps be fitted in conveniently with something else. Note that if a discussion period is planned part-way through the project, some of these activities should be undertaken before that takes place.

Some estimate of the part of the total time-budget needed for each step should be made, and then a full provisional timetable can be drawn up. Occasionally activities can go on in parallel – for example, if it is necessary to wait for printed computer output the introduction to the report could be begun while waiting. There will be any number of reasons why such a timetable will not be adhered to, but if something of the sort is drawn up consciously, at least at the beginning, it will help in the acquisition of good working habits (which are likely to be useful elsewhere).

In effect something rather like an elementary version of critical path analysis is being suggested, and while it would be ridiculous to treat this budgeting of time formally in such a way – indeed almost anything one says on the subject is likely to sound rather ponderous and pompous – it may be useful to draw the parallel.

Howard and Sharp (1983) describe a similar approach, but in considerably greater depth, since they are really concerned with a much more substantial project such as a PhD programme of research.

4.4.2 Report writing

It is perhaps not surprising to find that budding statisticians are not on the whole good at writing reports – after all, if this were their forte they would probably not be studying statistics – but any tendency towards such a shortcoming is certainly reinforced by the fact that, mathematics being largely written in symbols, mathematicians and other scientists, at least in the current British system, are not required to write much in the way of connected and cogent prose after the age of 16. Since, however, the great majority of graduates will have to communicate in writing during their working lives there is much sense in providing the opportunity for them to begin to acquire the rudiments of an acceptable style in the present context. In some programmes of study a separate item called, for example, communication skills may appear, and in this case the statisticians can devote less attention to discussing such matters (though it will be necessary to check that what is being taught elsewhere is consistent with what is expected for statistics).

Teachers, at least at higher levels, are reluctant to insist on a particular format for any kind of task, since they are aware that the format is to some extent a matter of taste, but for present purposes this is probably a mistake: the student in this context usually welcomes fairly specific guidance, and will be better able to develop his own approach, and to make his own judgement, when he has mastered the teacher's preferred layout. (In any case, if what is put forward by the student is as effective as, but somewhat different from, the model suggested, one can give credit in spite of the ignoring of instructions.)

Thus some instructions should be given on the structure of a report, and students should be told that correct grammar, punctuation, and spelling are expected, with an acceptable style; and that some credit will be available for this. This information is best put on paper, and a version used by ourselves is given in the Appendix.

As far as layout is concerned, it depends, at least in principle, on whom the report is aimed at, but little harm will be done if a uniform style is adopted, in so far as this is consistent with the different types of project. The following seems a useful structure.

First: a title page.

Second: a brief summary, never more than about half a page, and preferably considerably less, giving just the bare outlines of the contents, so that the (potential) reader knows what the report is about and can decide whether he will go further. In the present context, this summary could be little more than an extended title, and can normally be limited to a few lines, or a few sentences; a summary in the more usual sense is rather too long. The author's name should also appear, and the date.

Third: an introduction, describing the background and aims of the project, with a general indication of the methods adopted, and giving the conclusions. A reader should be able from this to get a good general idea of what has been done, so that, for example, he knows that subjects in a survey were selected as a stratified sample with proportional allocation.

Fourth: the *main* results and conclusions.

Fifth: as many sections as are necessary to give the details, the separate sections referring to the various stages, or distinct parts, of the investigation. Incidentally, so as to be able to decide to what extent the statistics should be explained in the report, the supposed reader needs defining. The most useful focus seems to be the student himself just before he began the project, i.e. a statistician well trained up to the same level as himself but unfamiliar with the details of how the project is to be done.

Sixth: a general discussion of the investigation and of the extent to which the investigation was successful. Any necessary reservations should normally appear here.

Seventh: references.

Eighth: appendices giving graphs, tables, computer programs, etc. (A point worth emphasizing at an early stage is that the submission of undigested computer output is unacceptable. Only the relevant parts should appear, and then should be annotated. The program/package used should be named, and the machine; a specially written program, unless it is so short as to be trivial, should probably be listed, but if so it must be accompanied by a flow chart or other key.)

Spelling is fairly rigidly specified in Standard English, but fortunately even the worst spellers (at this level) usually have trouble with only a few words. Bad grammar is less easily dealt with, but does tend to irritate those readers aware of it. Especially common errors are hanging participles ('evaluating the mean, the errors were found to be small' – this appears to say that the *errors* were evaluating the mean), and sentences without main clause verbs, or even any finite verb at all (though this is sometimes due to inappropriate punctuation). Punctuation problems usually amount to an absence, or at least a dearth, of punctuation marks; sometimes commas are used in place of full stops. Misuse of words, such as *infer* instead of *imply*, may also be mentioned here. Patient correction of the worst offences seems the only remedy.

Style is distinctly more difficult to deal with than the other aspects of writing, since reaction to it is so personal. A few general remarks are sensible: for example, to suggest that on the whole sentences should be kept short, and to remind the students that paragraphs break the material down into separate ideas, or into sections small enough to be dealt with comfortably. Logical sequence is not quite a part of style, but has the same property that perhaps the only test the writer can apply in the end is to read the report through, possibly aloud, after it is supposedly finished, as though it had been written by someone else. Conciseness is a cardinal virtue in this context, and methods for achieving it are arguably even more difficult to capture in simple rules, though preliminary thought about the main points to be conveyed and a readiness to cut out the superfluous can help.

One aspect of report-writing is the presentation of data (in tables, charts, etc.), and Chapman and Mahon (1986) offers much useful advice on this subject.

4.4.3 Oral presentation

There will probably be only small opportunity for an individual student to practise this, and perhaps in large classes none, but if it can be arranged it should be: perhaps one or two students can present short talks each week. It is practice which counts, for most people, but a number of general points can well be made explicitly to the students and the more common faults described – the need to plan the talk and if appropriate to prepare material in advance, and perhaps to try the talk out on a friend or on a mirror, the need to face the audience when speaking and to look at them, the need to speak appropriately loudly, the need for writing to be legible (in particular not too small, which is an especial danger with overhead transparency sheets prepared in advance), and finally not to go into too much detail. Mosteller (1980) and Freeman *et al.* (1983) are full of useful advice, of much broader application than just to the presentation of *scientific* papers; see also Chapman and Mahon (1986), Appendix A.

4.5 TYPES OF PROJECT

Of the various abilities listed and discussed in Chapter 2, different ones will be needed and, it is to be hoped, developed by different types of project: for example (see below) an *experiment* will, among others, test, at least to some extent, the ability to work as a member of a group (2.2.1a), while a *critical review* will not be concerned with that at all but rather with such things as communication (2.2.1c) and reading material critically (2.2.2g). There is no great value, we imagine, in an elaborate typology of projects, but it is worth discussing a few broad categories in order to show something of the various characteristics of projects. (A rather more detailed classification of the sample projects appears at the beginning of Section 5.2.) The development of new variants, and perhaps of major new types is, in any case, limited only by the imagination of the teacher and the availability of resources. The general aspects are considered in this section – a number of specific projects of various types are described and analysed in detail in the next section.

In some ways the most important single type is the *experiment*, or, somewhat similar, the *survey*. Only this allows group collaboration and close acquaintance with the investigation from the first stages of planning, through the data collection, to the final analysis and reporting. To be *told* that plans may be impracticable, or that data may be imperfect or incomplete or corrupted, is not as valuable as actually experiencing the difficulties (and so it is important to choose topics which are complicated enough to allow things to go wrong). One of the most interesting experiments (to us as teachers – whether the students felt the same is not so obvious) that we have carried out involved the growing of Chinese bean sprouts under various conditions of light, water, nutrient and so on; the only significant effect that appeared was the size of container (margarine tubs) used, which was not supposed to be a factor at all.

The extraction of information, both statistical and non-statistical, from another's writing, the re-expression of its salient features in an economical way, and an evaluation of its achievements, are skills not much exercised in the context of most experiments, however, and for these the writing of a *critical review* is useful. Often these will deal with broadly scientific subjects, but they need not; see, for example, Project 32 in Section 5.2.

One of our aims is certainly to argue that a wide variety of projects is necessary, in order to give wide experience; but fairly traditional activities are also valuable. A *statistical analysis* such as that invited in the French election question discussed in Section 4.6, or in the project about delinquency and family size (Section 4.2, Exhibit 2), is different, if at all, from commonly used material only because the translation into statistical terms is part of the problem, and because the method is not specified. Most such questions involve some model-building, which is one of their important features.

It is tempting, when writing or speaking of projects, to think only of rather grand themes which are suitable only for the more advanced students. But it is not difficult to create material appropriate at much lower levels, which can begin the process of enhancing some of the qualities desired. A requirement to find from the library the most recently published unemployment or overseas trade figures, and to compare them with those of another country requires students to *use* the library, and, perhaps with a little guidance, can lead to a realization that things with the same name may mean something quite different in other places or at other times. Similarly one can, at an early stage, stimulate the use of indexing and abstracting journals. Only someone who judges statistics by the weightiness of its theoretical content would regard these as valueless or unsuitable.

4.6 PARTICULAR PROJECTS DISCUSSED

In the rest of this chapter we discuss in more detail four specific projects suitable for use at various levels in a mini-projects course. The aim of the discussion here is to illustrate both the kind of question (open-ended, not technique-oriented) that is useful and the kind of response that such questions might elicit from the student. It shows too what in fact can be achieved by students, since many of the detailed points below have in fact arisen in students' project reports.

4.6.1 Social questions

The task for the student in this case is to extract information from a table or chart and to relate it to a statement, allegedly of fact, and then to discuss whether the statement is indeed justified; some technical statistical analysis may be called for, but typically nothing more than comparisons of percentages in different groups is really needed; some experience and general knowledge about society is also needed in most cases if the problem is to be tackled successfully. The title is not ideal, but has the advantage of brevity, and in any case, although the problem need not be concerned with social data, in practice it usually is, since social issues provide a readily available source of easily understood – or at least seemingly so – material and questions, in which sampling errors are not important and indeed often not present at all. We have found this type of problem very valuable in developing in students the twin abilities of putting their technical knowledge of statistical methods into perspective and of judging the extent to which legitimate deductions may be made from data. Somewhat similar tasks needing less general experience can be constructed; see, for example, Project 8, 'Delinquency and family size', in Section 5.2, which appeared earlier, as Exhibit 2.

A typical project, intended to represent a week's work, will have three or four separate parts, and one such part, drawn from one of the projects in Section 5.2 (Project 7), appears in Exhibit 4.

EXHIBIT 4

Life is getting easier for the criminal in this country: he is less likely to get caught and found guilty and, even if he is, his sentence will be lighter. (See Table 13.1 and Chart 13.9 of *Social Trends* No. 10, 1980.)
[The table and chart are reproduced with Project 7 in Section 5.2.]

As is usually the case the task is posed as a statement, of the sort that might be a newspaper headline or might be made in casual conversation by the man in the street. The tables towards which the student is directed are usually from official sources, and so presumably are reliable, but the conclusions contained in the statement may or may not be validly derived from the tables. There is sufficient connection, however, that one can see how a reader in a hurry might have been led to make the statement.

To some extent at least, one can develop a systematic approach to represent how an intelligent reader might analyse such a statement. (It is clear that to make sense of the problem the reader must be numerate, but it is scarcely necessary to be a statistician: indeed, as we observe below, it is quite easy for a statistician to go off in the wrong direction *because* of his technical knowledge.)

In this vein it is helpful for students to be reminded that the following questions – or something very similar – are implicit in any critical analysis:

1. What are the facts on which the statement is based?
2. Are they relevant facts?
3. What does the conclusion mean and does it follow from the data?
4. If the answer to the second question in 3 is in the negative, is there some other (closely related) statement that *is* true? Are there other data which do support the conclusion?

Although it is hardly necessary to discuss the specific problem in great detail here, it is worth showing how these questions can help to start the discussion off on the right foot. Experience shows, incidentally, that one can reasonably expect most students to get most of the points noted here.

The statement is presumably based on the facts that the ratio of 'people proceeded against' to 'serious offences' has declined over time, that the ratio of those 'found guilty' to those 'proceeded against' has also declined, and that the proportions of 'fines' and 'suspended sentences' have gone up while the proportion of 'imprisonments' has gone down.

These are in some general sense relevant, but, clearly, some reservations need to be made. First, the number of serious offences refers (it can only refer) to those recorded by the police: at different times and in different places many crimes are known to have been seriously under-reported, e.g. many kinds of assault, including rape, and increasing reporting rates over time would distort the picture.

Similarly, decisions to begin proceedings certainly depend on policy, which may have changed over time. Lastly, it seems that those found guilty of serious offences in England and Wales in 1951 numbered 133,000, which is more than those proceeded against (114,000); the explanation almost certainly lies in the time delay, first between the offence and the laying of charges, and second between the laying of charges and the verdict being given. Thus the various figures for any given year are only roughly comparable.

Do the conclusions follow from the data? In view of the distinction between 'crimes committed' and 'crimes reported' and of the time delays it is impossible to be sure of what lies behind the declining ratio of 'people proceeded against' to 'serious offences', or indeed the declining ratio of 'people found guilty' to 'serious offences'. The ratio has not changed much in England and Wales in any case, and Northern Ireland always gives rise to particular difficulties because of the political situation; but the change in the ratio for Scotland – it has roughly halved – is perhaps large enough for one to conclude that in that part of the kingdom the chance of proceedings being brought in respect of a crime has decreased. (More than one person can be involved in a crime, and a change in the distribution of crimes among types might because of this distort the picture, but like other complicating factors it seems unlikely to be sufficient to account for the magnitude of the change.) As for the question of whether a criminal is less likely to be found guilty if he is caught, there is of course no evidence at all in these data: there is no reason to suppose that everyone who appears in court – or even a fixed proportion of them – is in fact a criminal. In considering the question of severity of sentence two points ought to be made: any comparison of the severity of the various types of sentence is at least partly subjective; and, in any case, changes in the sentences imposed can only usefully be discussed crime by crime – conditionally in fact – whereas the tables provide only marginal information, and a change in the distribution of the type of crime committed over the years would very likely also change the distribution of sentences imposed. There does not seem much more information to be got out of these data.

There are two noteworthy errors into which students often fall in answering this kind of question. One is to mix beliefs and prejudices with statements based directly on data, often without realizing it: it is perfectly acceptable to observe that 'it is widely agreed that changes in penal policy have led to a replacement of prison sentences by fines', or that 'it is well known that the police no longer try very hard to clear up burglaries in the big cities', but it is not legitimate to use such statements – unless one can give a reference to hard evidence on the question – as though they have the same weight as conclusions based on the data themselves. The other error is to apply statistical techniques without any clear justification: for example, to apply a χ^2 test to the numbers of persons proceeded against and the number of offences reported, in order to test whether the ratio has changed, even though it would be most natural to regard these as including the whole population rather than a sample.

4.6.2 Experiment

We take as an example of this kind of assignment a project in which the students are asked to plan and carry out (collaboratively) and report (individually) on the results from a small-scale agricultural/horticultural experiment. The main aims are to give first-hand experience of some of the practical considerations in planning experiments and an appreciation of the care needed in carrying them out. Subsidiary aims are to give practice at working in a group – at agreeing objectives and setting up an organization to achieve them – and to give experience in analysing data whose origin is known in intimate detail. Students who have already met the elementary ideas of experimental design and analysis are likely to gain most benefit from such a project.

There is of course no reason why the subject matter should be confined to the agricultural or horticultural area, though experiments with plants are certainly convenient and produce many benefits from the pedagogical point of view – such as the unpredictability of the course of an experiment. For groups of students with special interests in some area, it may however be possible to choose more directly relevant experiments. What *is* important, no matter what the subject, is that the students should see the outcome as useful or interesting, or at least as similar to something which is useful or interesting (the mustard-and-cress experiment of the next paragraph may not be directly useful, but obviously experiments of a very similar type on other material are).

The specific project discussed here is an investigation of the best method of growing mustard and cress in the home; this appears as Project 17 in Section 5.2. Mustard and cress is a useful subject because it grows from seed to maturity within 10 to 20 days and so the whole project may be comfortably fitted into 7 or 8 weeks, thus allowing 2 to 3 weeks for planning before the experimental phase and a week or two for analysis of the results afterwards. For much of the duration of the project it makes relatively small demands on the students' time, and so may be treated for all except, say, its final two or three weeks, as a 'background job', work on which continues at a low intensity in parallel with more pressing projects.

EXHIBIT 5

At a later stage you will be asked to carry out an experiment concerning the growing of mustard and cress in the home, in order to investigate the effect on growth of such factors as the growing medium used, the amounts of light and nutrient provided, competition with the other crop, and any other conditions that you think might be of interest. Seeds, seed trays, plant nutrient, and various types of growing medium will be available.

At this stage in the project you are asked to draw up plans for the experiment, deciding first on the particular aspect(s) of growth which you wish to investigate and secondly on the various treatments you will use. The plans will need to include, among other things, details of the experimental design that will be used, the conduct of the experiment, and the precise requirements for equipment and materials.

The first phase of the project is the introductory and planning phase, which might be initiated by issuing to the class a project-sheet as shown in Exhibit 5.

It will probably be helpful to supplement this with an introduction from the teacher in which he describes more fully the background to the project and indicates in broad terms what kinds of resources and materials could be made available, and in what quantities. For the background, he could specify the general aims of such a project, as set out above, and then, for the particular experiment in question, suggest that, although the instructions printed on the seed packets claim that mustard and cress will grow on, say, cotton-wool or tissue paper, limited domestic experience has suggested that better results are obtained from other materials such as seed-compost, and that other factors, such as light and heat, appear to influence the quality and yield of the crop. On the question of materials he could, for example, draw the class's attention to the easy availability of cheap plastic seed pots of various sizes and tell them that, as a start, if required, growing media such as vermiculite, cotton-wool, tissue-paper, compost, etc., and various plant foods are all easily obtainable. A conscious attempt by the teacher to appear amateurish about horticultural matters at this stage will help reduce the danger of the class's attempting to discern in these suggestions some ready-made 'right answer' to their design problem and of them wasting effort in trying to extract it from him. The teacher's aim should be merely to provide a realistic starting point for a wider discussion of the real problem.

The discussion of statistical plans for the experiment is perhaps best left for a subsequent meeting, when members have had time to reflect on the problems and gather further information that they think relevant. The latter might well include the fact that mustard and cress are distinct plants, which grow at different rates, and so it might be sensible to restrict the experiment to the faster-growing mustard. Information on the size of seed, etc., might also be helpful. The fact that the question is concerned with growing mustard (and perhaps cress) *in the home* will no doubt be noted somewhere, preferably by the students: elaborate lighting or watering regimes are therefore not appropriate, for example. For the conduct of the discussion many of the general comments on discussion groups in Chapter 3 are relevant. In particular it will lead to most learning if the teacher keeps as much as possible in the background, limiting his participation to the asking of disingenuous questions where appropriate and the supplying, if he has it, of relevant background information. As suggested above, he should try to keep his green fingers, if he has any, out of sight.

To focus the discussions, however, the two questions: (i) 'What factors do we wish to investigate?', and (ii) 'How shall we design the experiment in order to investigate them?' will have to be raised. Tentative answers to (i) can lead to (ii), in the light of answers to which (i) may need to be revised and the process repeated iteratively. In our experience initial answers to (i) are often over-ambitious and consideration of (ii) can have a salutary sobering effect. Some preliminary reading by the students on considerations to bear in mind when planning an experiment might be beneficial. We have found, for example, a short paper by Finney (1982), noting questions to ask in planning a comparative experiment, particularly useful in this respect. Some questions relevant to the present problem suggested by Finney's paper are:

What are the aims?
  Which variate will be measured?
  Is the aim to detect significant differences (for example between growing media) or to estimate numerical quantities?
Which treatments will be used?
  How are they structured? Is there any pattern or hierarchy?
  Is there a linear structure—if so, what levels?
  Is there a factorial structure?
What experimental unit will be used?
  What is known about variability between units?
What resources are available?
  Are there constraints on their use?
  How many units can be used?
Conducting the experiment and recording the results:
  How, and by whom, will the experiment be carried out?
  How will the results be measured, and how recorded?

If the students have read Finney's paper or have some similar background, many of these questions will occur naturally: if not, the teacher might feel it desirable to introduce some of them as occasion demands. His aim should be to encourage (a) recognition of the factors affecting choice of a design, and (b) the subjecting of the various possibilities generated in the discussion to careful rational assessment.

The outcome of such discussions is, of course, unpredictable, and so the present account continues with a description of one particular design and subsequent experiment adopted by a group of seven Sheffield students in the Spring of 1983. Two discussion meetings of about $1\frac{1}{4}$ hours each were held. At the first, various factors potentially of importance and interest were identified. These included: the effect of density of sowing, the effect of watering, and the influence of light on both germination and subsequent growth, as well as factors mentioned in the teacher's introduction to the project. It was decided that experimental units should be small plastic seed pots of size 6.5×5 cm readily available at a local garden shop. It was recognized that the quality of a crop of mustard was not just a matter of its size, as measured, say, by weight, but also depended on its appearance and taste. One important piece of information necessary for detailed planning was found not to be known by anyone: the amount of seed needed for a reasonable density of sowing for a single seed pot. The group undertook to discover (by some simple weighings – on a balance capable of measuring to 0.1 g) this value before the next meeting. (In a different year another group of students carried out a simple pilot experiment to discover roughly how much seed was needed and how much water was appropriate.)

At the next meeting, about one week later, final decisions were taken on the treatments to be used, measurements to be made and on the design of the experiment. It was decided to use the following treatments with levels indicated:

Growing medium	compost
	vermiculite
	tissue paper
Light during germination	cover during germination
	do not cover during germination

Density of sowing $\left\{\begin{array}{l} 3\frac{3}{4}\,\text{g} \\ 2\frac{1}{2}\,\text{g} \\ 1\frac{1}{4}\,\text{g} \end{array}\right.$ per pot

Measurements were to be:

Weight of crop

Edibility – a subjective assessment on a three-category scale, consisting of 'appealing', 'tolerable', and 'repulsive'.

It was recognized that the amount of water given to each pot was likely to have a very large influence on the measurements, and so, in view of the aim of the experiment to provide guidance on the best growing conditions in the home, it was decided to give a standard amount of water for each medium whenever required to keep it moist; this was what was advised on the seed packet instructions. The possibility of estimating any effect due to water separately from that due to the other factors was therefore deliberately sacrificed.

To limit cost and experimental effort it was decided to use no more than 60 experimental units, and eventually a design consisting of three replicates of a $3 \times 2 \times 3$ factorial arrangement was adopted. The site for the experiment was to be a large bench in front of a high east-facing window. The group thought that distance from the window might affect growth, and so finally decided to lay out replicates parallel to the window, in three rows of 18 pots each, one behind another. The replicates were thus turned into blocks. Treatment-combinations within each block were allocated randomly. The details of the design, including practical aspects such as the intended watering regime and the method of sowing and harvesting, were required to be recorded before proceeding, in an experimental protocol, giving enough information to enable someone else to carry out the same experiment. Sowing took place a few days later, to give time for assembly of all the materials and equipment required. The group organized a duty rota for care of the emerging seedlings. It was found that daily attention was needed to avoid difficulties with drying out. After 14 days the crop was harvested and measured. The timing here was chosen for convenience to fit in with the regular weekly meetings of the class, but a shorter or longer growing time would have been feasible, since the mustard plants appeared to reach maturity after about 12 days and would have survived for some time after that before deteriorating. Although the particular experiment described here ran reasonably smoothly, in other similar experiments various accidents have occurred (such as the overturning and destruction by an over-zealous cleaner of part of an elaborately balanced design and the demise in another of many of the seedlings because of neglect over a critical weekend) and these have brought home quite forcefully the desirability of careful experimental technique and the difficulty of analysing experiments with missing data. (They also suggest a novel meaning of the term 'robust' as applied to experimental designs.)

With the harvest and measurement the experimental phase of the project ended, and the analysis and preparation of a report began.

4.6.3 Critical review

An example of this kind of project is shown in Exhibit 6 (Project 23 of Section 5.2).

EXHIBIT 6

WEATHER FORECASTING

The references below are concerned with different aspects of the problem of monthly weather forecasting.

Write a critical review of each paper, explaining:

  (i) the purpose of the paper;
 (ii) its assumptions, method, and conclusions;
and including, if appropriate,
(iii) a discussion of any limitations; and
(iv) any further analysis that seems desirable to you.

Comment, in the light of your reviews, on the state of monthly weather forecasting as reflected by these papers.

A. GORDON (1974) Accuracy of weather forecasts, *Nature*, **252**, 294–295.
F. H. W. GREEN (1975) The February–June weather relationship in north-west Europe, *Nature*, **253**, 522–523.

For convenience the two papers referred to in Exhibit 6 are reproduced in Exhibits 7 and 8, by kind permission of the Editor of *Nature*.

EXHIBIT 7

ACCURACY OF WEATHER FORECASTS

After nine years of publication there has been considerable criticism of the usefulness and accuracy of long range weather forecasts. Gordon[1] has suggested that the forecasts have not yet attained a practical or economical degree of usefulness, and that advance computer technology and statistical power will not improve their accuracy unless a breakthrough in ideas is achieved. Existing methods are broadly based on an analogue or matching technique in which overall weather circulations of a particular month for many decades in the past are compared with those of the month preceding that for which the forecast is to be made.

It is, therefore, clearly necessary that various assessments of the results are made independently of those calculated by the predictors.

I have devised a classification to evaluate the accuracy of the predictions for the London area, from observations made at Kew Observatory. The portion of the long range forecast used is that part which refers to southern or south-eastern England. Although the forecasts mention several parameters, I have selected temperature and rainfall for verification (probably the two most important factors for the consumer, whether in the public or private sector of the national economy). This study deals with the 108 months from June 1965 to May 1974, inclusive.

Table 1 Assessment of official long range forecasts

Predicted	Observed	Good	Bad	Very bad	Actual distribution
Above average	Temperature	15	10	10	42
	Rainfall	3	5	9	31
Average	Temperature	6	32		22
	Rainfall	22	36		38
Below average	Temperature	16	6	13	44
	Rainfall	13	11	9	39

Table 2 Scoring in systems A and B

All forecast above or below average		All forecast average
$2 \times 3 = 6$		$2 \times -2 = -4$
$1 \times -2 = -2$	(assuming the expected distri-	$1 \times 8 = 8$
A	bution of 2, 1, 2 in 5 trials)	
$2 \times -2 = -4$		$2 \times -2 = -4$
Total 0		0
$2 \times 4 = 8$		$2 \times -2 = -4$
B $\;\;1 \times -2 = -2$		$1 \times 8 = 8$
$2 \times -3 = -6$		$2 \times -2 = -4$
Total 0		0

The forecasts predict mainly one of three probabilities, below average, average (which includes near average or about average), and above average. A few forecasts of temperature have predicted much above average or much below average but these classifications have been treated simply as above average or below average in this study.

The boundaries which separate the classes above average, average, and below average are not published with the forecasts in the press or in the Weather Log (in *Weather*) and in many cases the public must make an inspired personal judgment. The Meteorological Office will, however, supply this information on request and it is understood that subscribers to the long range forecasts are provided with the exact boundaries. These are quintiles for temperature and terciles for rainfall, and are based on the climatic mean for the particular month and area of the United Kingdom for which the forecast applies. In the case of rainfall, $\pm 20\%$ of the mean climatic total defines the middle tercile or band width for the average condition. In the case of temperature the middle quintile for Kew is about $\pm 0.35°C$ when averaged throughout the year; it is used in the forecast assessment to define the average condition.

Table 3 Sequential annual point totals 1965–74, inclusive (108 cases)

Year	Temperature		Rainfall	
	A	B	A	B
1965*	1	3	−1	−1
1966	16	19	3	2
1967	−4	−2	−3	−4
1968	−9	−10	−3	−4
1969	6	8	3	5
1970	21	27	3	1
1971	−14	−17	0	0
1972	−9	−11	3	5
1973	−9	−10	3	5
1974†	0	0	−2	−5
Total	−1	7	6	4

* 7 months.
† 5 months.

The classification system used here to assess the forecast results considers nine boxes, three for each class of prediction. If the observed result falls in the same box as the predicted result the forecast is 'good', if in a box adjacent to the predicted result the forecast is 'bad', and if in a box two boxes away from the predicted result the forecast is 'very bad'. In the last case it means that a forecast for above or below average conditions has been made whereas the observed result has been the opposite, that is, below or above average, respectively.

Table 1 shows the results for temperature and rainfall, together with the actual distribution of observed results. The observed quintile distribution follows closely the ratios 2,1,2 where much above and much below average are included with the above and below average cases. The observed tercile distribution follows the 1,1,1 ratios acceptably well for the rainfall.

Two scoring systems may be used. The first is that applicable to a game of chance. The second is rather better for the purpose of testing a scientific prediction technique in that it gives a one point bonus for a correct above average or below average forecast and a one point penalty for a very bad forecast as already defined.

For temperature, the quintile distribution is divided into the proportions 2,1,2. The probabilities for each class above average, average and below average, respectively, are therefore 2/5, 1/5, and 2/5. These three probabilities are equivalent to odds of 1.5 to 1, 4 to 1, and 1.5 to 1, respectively (3 to 2, 8 to 2, and 3 to 2, respectively). The gains for a success are therefore 3 for above average, 8 for an average and 3 for a below average. A failure in any class loses 2 points and this is indicated by − 2. That explains the scoring system if the rules of a game of chance are operated (system A).

Scoring system B is exactly the same as system A except for a modification which allows arbitrarily one point extra if a success is forecast for above average or below average. Thus, a success for above average is 4, for average 8 as before, and for below average 4. To equalise this a penalty of one point is invoked if the result is the opposite of the forecast of condition. For the latter case the loss is -3 instead of -2.

Both systems add up to zero if all cases are forecast in any one category (Table 2). The total scores for the nine-year period studied are:

Temperature: system A, -1; system B, 7. Rainfall: system A, 6; system B, 4.

None of the scores approaches an acceptable level of statistical significance as given by the chi square test for 108 cases. Wright and Flood[2] found no statistical significance for the temperature and rainfall forecasts in an assessment covering the results of forecasts for the whole country for a two-year period. Some statistical significance was, however, found for the general information forecasts. The latter comprise statements of the kind "soon becoming wetter", "dry weather at first", "some warm days", and so on. In view of the fact that such statements are often very subjective in character it is extremely difficult to assess the results quantitatively. For this reason, and because the predictions are often of a synoptic scale type and, therefore, not really long range forecasts, it is doubtful if the significance is meaningful in the context of a 30-d-period.

The results of the long range forecasts for a month ahead for temperature and rainfall do not approach an acceptable level of statistical significance. In fact, it is doubtful whether one can honestly say that they are even marginally better than chance. Although the selected cases only represent a sample of the total population of cases which could include a further 10 areas, the London area could be considered as one of the most important for verification. Further studies on the other areas should be done but it is doubtful if the net result would show any marked improvement. Furthermore, it is extremely doubtful that Ratcliffe's claim[3] of a modest but real improvement in the last few years in the long range forecasts is now valid. Table 3 shows the year-by-year scores for systems A and B for both temperature and rainfall.

It is also questionable whether the band width of $\pm0.35°C$ for the middle quintile for temperature is at all meaningful as a definition of average temperature for the consumer. A band width of $\pm1°C$ might be far more useful to the suppliers of gas, electricity, coal and so on, for planning purposes.

It is certainly time to consider whether the very considerable expense incurred by this branch of our National Meteorological Service should continue to support what might be judged as a profitless activity. Bearing in mind that it is recognised that there is a built-in persistence factor in the sequence of meteorological conditions, almost any person conscious of the weather should be able to obtain a small positive score after a sufficient number of trials, given the previous months anomalies.

A. GORDON

14 Linkway,
Edgcumbe Park,
Crowthorne,
Berkshire, UK

Received May 9; revised July 2, 1974.

[1] Gordon, A. H., *Weather*, **28** (6), 264 (1973).
[2] Wright, P. B., and Flood, C. R., *Weather*, **28** (5), 178–187 (1973).
[3] Ratcliffe, R. A. S., *Met. Mag.*, **99** 125–130 (1970).

EXHIBIT 8

THE FEBRUARY–JUNE WEATHER RELATIONSHIP IN
NORTH-WEST EUROPE

It has been well said by Bonacina[1] that everything which happens to, or in, the atmosphere affects its subsequent behaviour. In other words, there is a "memory" in the atmosphere, probably such that an anomaly produced at one time may lead to a similar, or a related, anomaly being restored in the future. It is known that although some factors can be neglected for short range forecasting, they become progressively more important in the longer range. These factors are unknown, although most meteorologists would probably agree with Namias[2] that two of the most important must be extraterrestrial events, such as variations in solar activity, and variations in the character of the Earth's surface.

But empirical methods of long range forecasting are sometimes found simply by discovering relationships between observations at different times and places, and there seems no reason for not using them in prediction. The physical mechanism behind the relationship may not be fully understood, but the empirical discovery usually points to the research required for understanding the mechanism. Here I illustrate this idea by an example.

In the course of looking at climatic trends in north-west Europe, particularly the northern North Sea area, certain relationships were noted between weekly or monthly periods in the winter season and periods of similar length in the following summer. This may best be illustrated by what one may call the February–June relationship. Figure 1 shows the relationship between February and June from 1940 onwards for Dalen i Telemark (59°27′N, 8°0′E), an inland station in southern Norway. A regression line can be drawn which shows that the mean temperature in

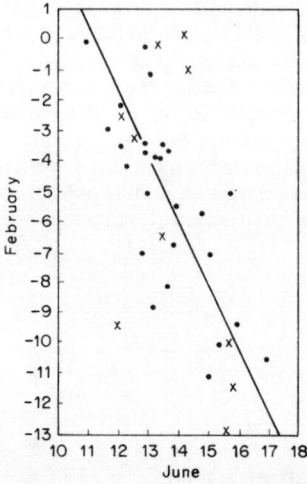

Fig. 1 February and June mean temperatures, °C, Dalen i Telemark, 1940
onwards. (Pre-1950 shown by crosses.)

Table 1 Dalen i Telemark, mean temperature in June (°C)

	Forecast	Actual	Trend from previous June correct (C)	Forecast within 1°C of actual (C)
1939	—	13.6	—	—
1940	16.5	15.8	C	C
1941	15.9	15.7	C*	C
1942	15.6	12.0	C	
1943	11.3	13.4		
1944	12.4	12.2	C	C
1945	12.9	13.2	C	C
1946	12.8	12.7	C	C
1947	17.3	15.5	C	
1948	14.2	13.5	C	C
1949	11.1	14.3		
1950	13.0	13.0	C	C
1951	13.2	12.9	C*	C
1952	13.0	12.6	C*	C
1953	13.8	15.9	C	
1954	14.3	13.3	C	C
1955	14.4	12.9		
1956	15.0	14.3	C	C
1957	13.3	13.4	C	C
1958	14.8	13.7	C	
1959	13.8	14.1	C	C
1960	14.4	15.1	C	C
1961	13.3	13.8	C	C
1962	12.8	12.1	C	C
1963	15.0	15.4	C	C
1964	13.2	12.2	C	C
1965	12.9	13.3	C	C
1966	15.6	15.1	C	C
1967	13.2	13.6	C	C
1968	13.9	14.9	C	C
1969	15.1	16.0	C	C
1970	15.5	16.8		
1971	12.5	13.2	C	C
1972	13.4	12.4		C
1973	12.5	14.4	C	
1974	12.6	13.9	C	

* Almost the same as previous year.

February is inversely related to the mean temperature in June. If this regression line is used at the end of February to estimate the mean temperature for the following June, the results compare very favourably with the actual temperatures (Table 1).

It will be seen that in 30 years out of 35 the estimate was correct in the sense that it predicted correctly whether the June mean temperature would be higher or lower than (or very nearly the same as) that of the previous year. In 25 years out of 35 the mean temperature was predicted correctly to within 1.0°C and in 14 of those years it was correct to within 0.5°C. The accuracy of the estimates increased until the past few years, when it has deteriorated slightly. Some variations in this simple predictive method were investigated, mainly to try to take into account year-to-year trends, but this has not led to any general improvement in the method.

Periods other than calendar months were not used because of the greater amount of work involved. Had this been done, it would have rectified some, if not all, of the least successful predictions. For example, the prediction of a very warm June in 1955 proved to be incorrect, but in fact July proved to be exceptionally hot.

This February–June relationship was explored (1) backwards in time, and (2) in neighbouring areas. At Dalen i Telemark, during most of the 1940s, the relationship came close to the regression, but this was not the case from 1891 to the 1940s. There was, however, a suggestion of two populations, one with a direct correlation and another with an inverse correlation, the latter applying to all cases where the February mean temperature was −6.0°C or below. Examination of other stations in Norway, Denmark, Scotland and England indicated that the inverse correlation between February and June mean temperatures had been spreading south and west. There is more scatter about the regression line at coastal stations, but, for practical purposes this is compensated for by the angle of slope of the regression line approaching 45°.

Table 2 Bergen (Frederiksberg) rainfall in June (mm)

	Forecast	Actual	Trend correct compared with previous year (C)	Forecast correct within 50 mm (C)
1950	160	203	—	C
1951	163	72	C	
1952	140	225	C	
1953	132	46	C	
1954	115	137	C	C
1955	95	57	C	C
1956	90	165	C	
1957	150	112	C	C
1958	100	26	C	
1959	155	182	C	C
1960	135	166	C	C
1961	180	223	C	C
1962	155	116	C	C
1963	80	68	C	C
1964	153	252	C	
1965	147	187	C	C
1966	105	106	C	C
1967	177	163	C	C
1968	125	132	C	C
1969	80	80	C	C
1970	100	78	C*	C
1971	175	69		
1972	153	247	C	
1973	157	130	C	C
1974	195	131	C	

*Actual precipitation almost the same as previous year.

Table 3 Kinlochewe, June rainfall (mm)

	Forecast		Actual	Trend correct compared with previous year (C)		Forecast correct within 50 mm (C), within 25 mm (C*)	
	X	Y		X	Y	X	Y
1954	110	112	123	—	—	C*	C*
1955	125	102	82		C	C	C*
1956	120	75	75		C	C	C*
1957	140	145	82	C	C		
1958	135	82	60		C		C*
1959	170	125	201	C	C	C	
1960	110	170	190	C	C		C*
1961	175	147	137	C	C	C	C*
1962	165	170	190	C	C	C*	C*
1963	105	70	76	C	C	C	C*
1964	165	150	103	C	C		C
1965	145	165	164	C	C	C* ·	C*
1966	135	45	149	C	C	C*	
1967	175	147	124		C		C*
1968	110	120	120	C	C	C*	C*
1969	105	70	45	C	C		C*
1970	110	50	66	C	C	C	C*
1971	170	182	94	C	C		
1972	140	140	150	C	C	C*	C*
1973	140	182	177		C	C	C*
1974	165	180	70	C			

X—using Kinlochewe February mean temperature.
Y—using Dalen i Telemark February mean temperature.

In both Norway and Scotland, higher than average temperatures in winter correlate well with higher than average rainfall, since both occur in westerly and cyclonic conditions. Conversely, there is relatively little precipitation in anticyclonic conditions. In summer there is an inverse relationship between temperature and rainfall. Thus forecasting temperature is almost tantamount to forecasting approximate rainfall. In the case of Bergen, Norway, in 22 consecutive years out of 23, June rainfalls were correctly predicted at least to the extent of being greater or less than in the preceding year, and in 13 years out of 25 the amount was predicted to within 25% (Table 2).

Mean February temperatures from Dalen i Telemark were used to predict mean June temperatures and rainfall from 1954 onwards for Kinlochewe, in Wester Ross. For temperature, the predictions were nearly as good as when February temperatures for Kinlochewe were used, but for rainfall they were usually better (Table 3). The particular case of 1974 is interesting. February was mild, and followed a quite exceptionally mild January. The predicted rainfall for June was 165 mm. In the event, the June rainfall was only 69.6 mm at Kinlochewe, but there was a very marked spell of wet weather, preceded and followed by dry conditions, commencing on May 18; the rainfall for this period was 164.6 mm, suggesting that the very mild and wet period in January–February was preconditioning a wet spell in May–June.

The meaning and physical explanation of this empirical prediction technique can only be guessed at. The second of Namias' factors—variations in the character of the land or sea surface—may be a more direct key to the relationships described than is the extraterrestrial factor. Can one determine, for instance, what was the particular state of which part of the Earth's surface in successive Februaries which injected a factor into the atmospheric circulation that was effective in inducing another particular condition the following June? It is not unreasonable to suppose that the surface conditions in February reflect the dominating influence which has characterised the whole winter.

Two significant factors which come to mind are (1) the generally accepted spread southwards of the Arctic high pressure area, and (2) the related decrease in westerly days in the zone to the south of this. It is significant that Trondheim, the northernmost station considered, showed the inverse February–June correlation earliest, and that it became recognisable further south and west as time went on. This implies increasing anticyclonic control of the relationship. Markedly anticyclonic and cold Februaries in Norway have almost invariably preceded warm Junes, and since well before the 1940s; but in addition it seems that milder, westerly Februaries now precede cool wet Junes more regularly than they used to.

In view of the very limited success of the methods at present used by the Meteorological Office in producing monthly weather outlooks[3] I feel that studies of the type described here could be very useful ingredients in the development of long-range forecasting technique, especially when correlated with the *Grosswetter Lager* and other atmospheric circulation studies. I am pursuing this matter, with some success, in cases other than the February–June relationship.

I thank Dr E. M. Frisby for many suggestions during the course of this study, and the Norwegian Meteorological Institute for the provision of observational data in advance of publication.

F. H. W. GREEN

Department of Agricultural Science,
University of Oxford,
Parks Road, Oxford, UK

Received November 22, 1974.

[1] Bonacina, L. C. W., *Weather*, **28**, 382–385 (1973).
[2] Namias, J., *Unesco Courier*, Aug./Sept. (1973).
[3] Gordon, A., *Nature*, **252**, 294–296 (1974).

The breakdown of the task into four parts is intended to define more precisely for anyone attempting the project what is meant by a 'critical review'. The intended meaning is evidently closer to literary criticism than to an invitation merely to point out shortcomings. A critical review of a paper is an interpretation and evaluation which should be helpful to a prospective reader, enabling him to see quickly in advance what the paper is about, how, in outline, it is constructed, and to what extent it is successful. It will be helpful for many students setting out to write such a review to suggest that they try to keep in mind an imaginary reader who may be assumed to be scientifically educated and statistically aware but not so highly trained as themselves, and who has no special knowledge of the subject matter of the paper. A student a few months behind them on their own course, or a younger version of themselves, might for example give the right kind of focus for writing. With such a reader in mind the report will need to clarify obscurities, and explain novel methods (if the explanation in the paper itself is likely to give pause to such a reader), but will not go into detail about standard techniques.

To illustrate the type of explanation and comment that the review might usefully include we look in more detail at the present example on long-range weather forecasting. The two papers discuss different aspects of forecasting: Gordon uses a scoring system to assess the accuracy of UK monthly forecasts issued between 1965 and 1974, and Green discusses apparent connections between winter weather in southern Norway and the weather during the following summer in the same region and in other areas bordering the northern North Sea; he suggests that such relationships might be useful ingredients in future techniques for long-range forecasts.

The four questions posed on the project-sheet might be used as a framework for the writing of a review. When, as in the present case, more than one paper is being reviewed it will be a matter of taste and of judgement of the detailed contents whether separate reviews along such lines for each paper or a single consolidated report is better. The former is probably simpler in general, and usually adequate if supplemented by a final section containing comments on relationships between the papers, and drawing any overall conclusions.

Accordingly we consider the two papers separately, listing brief comments under these four headings.

'Accuracy of Weather Forecasts' – A. Gordon

(i) Purpose.   This has been briefly described above, and perhaps only a small amount of further detail – on, say, the forecasting method used – needs to be added. In general, though, this section of a review might contain also some indication of why the topic of the paper was of interest, if that is not self-evident. Students might find it helpful in their search for the material to include in this section to be reminded of the loose pattern to which many technical papers roughly conform: like the ideal lecturer, they say what they are going to say, then they say it and then they say what they have said; that is, they contain a general introductory section, which will usually contain the material for this part of the review, followed by one or more more detailed technical sections and then, at the end, a more general discussion of results and summary of conclusions. Recognition of such conventions, which to the experienced reader are so familiar that they are almost part of his subconscious, can provide useful orientation for the novice.

(ii) Methods and conclusions.   Any description of the detailed argument of the paper will no doubt concentrate mainly on the scoring systems used, and the rationale of the tests based on them. Points that probably deserve particular explanation are:

(a) the mechanics of forecasts: the meaning and role of quintiles and terciles;

(b) the justification for the specific scoring systems;

(c) details – not spelled out in the paper – of the formal tests based on the scores. What, for example, was the null hypothesis? Students may here find a normal test more natural than the χ^2 test mentioned in the paper, if, indeed, they can discern precisely what is meant by the χ^2 test.

This part of the review will probably be found quite challenging, since the paper is both reticent on some details and is written in terms that to a statistician many seem rather indirect. The main conclusion however is unmistakable: 'it is doubtful whether one can honestly say that [the forecasts] are even marginally better than chance'.

(iii) Limitations.   Perhaps the major limitation is the arbitrariness of the scorings.

(iv) Further analysis.   Two possibilities are:

(a) An investigation of other scoring systems. How far, for example, would the proposed scoring systems need to be changed in order to alter the paper's conclusions?

(b) An analysis of the data without the use of a scoring system. It is possible, for example, to reconstruct from Table 1 of the paper the two-way classification of the 108 forecasts according to predicted level (above, at, and below average) and realized levels. Techniques for the analysis of categorical data can then be applied.

'The February–June Weather Relationship in North-West Europe' – F. H. W. Green

(i) Purpose.   Illustration of the existence of such relationships by examples of February and June weather from places in Norway and Scotland.

(ii) Methods and conclusions.   The argument of the paper rests mainly on simple linear regression and a discussion of both the sizes of residuals (corresponding to a forecast being within a specified distance of the actual value) and whether a fitted value lies on the same side of the previous year's observation as the actual value (corresponding to a correctly 'forecast' year-on-year trend). The review would clearly need to spell out these arguments in some detail.

(iii) Limitations.   Among points that can be made are:

(a) The discussion in the paper is entirely in terms of relationships fitted to the complete data, so that 'forecasts' are not estimates based on past values, as genuine forecasts would need to be, but are merely internal fitted values.

(b) There is some suggestion in the paper that the illustrations presented had been selected from a much larger mass of data going back further in time and relating to other places. For one of the sites (Dalen in Norway) the apparent relationship is claimed to hold over the years from 1940 to 1974: earlier data between 1891 to 1940 show no such relationship. There is a possibility therefore that selection effects have some part to play in explaining the apparent relationships: given a large enough body of *independent* data it will be possible to find subsets which by chance alone show a high correlation. At the very least therefore the possible presence of such effects makes it difficult to judge the strength of the evidence for the claimed relationships. This point was made quite forcefully in a letter to *Nature* on 25 September 1975, by I. Kanestrøm of Oslo. He showed that over the years 1890–1972 there was no significant correlation between February and June temperatures in Dalen, and that only by considering a shorter period back from 1972 could a significant correlation be found. On the basis of this and similar evidence from elsewhere he concluded that 'there is no statistical evidence that winter temperature is a useful element in forecasting the summer temperature and rainfall for the same domain'.

(c) The scatter plot for the first example in the paper – the only one for which a plot is shown – suggests that the slope of the regression line is strongly influenced by a few extreme points: without these the line would be much more nearly horizontal. The evidence for a relationship therefore in this case at least rests rather heavily on extreme temperatures in only a few (5 or 6) years. Kanestrøm makes this point too.

(d) The points on the scatter plot do not seem to agree entirely with the data in the table, as one of our students discovered.

(iv) Further analysis. In the absence of new data there seems little scope for further analysis, though an exceptionally enterprising student might be tempted to speculate about the effect on the paper's argument of year-to-year dependencies or non-stationarities in the data.

These are some of the detailed points that might be included in a critical review. We would not expect even a good student necessarily to mention all of them, but we would hope that most students would be able to describe the overall purpose and conclusions of the papers reasonably clearly and to offer some explanation – going beyond that in the papers – of the arguments used. On the final requirement of the project-sheet – to comment on the state of monthly weather forecasting, or, in other words, to attempt some overall conclusion – we would expect some expression of scepticism about the effectiveness of the monthly forecasts available at the time of writing of the papers, and about the prospects of using the kind of relationships described in the second paper for improving them. Students, in our experience, do usually manage to rise to these expectations, and some produce penetrating and reasonable criticisms going well beyond them.

The benefits that a student might derive from working on a project of this type can include:

- the development of an approach of his own to the extraction of information from a paper, and, in particular, practice at taking an overview of a piece of work;
- the gaining of confidence in approaching the literature on other topics;
- practice in the understanding and evaluation of technical arguments applied to a substantive problem;
- the stimulus to think for himself about the problem addressed in the paper, and perhaps the stimulus to develop his own approach to it;
- practice at explaining obscure or unfamiliar technical matters, and, more generally, at expressing (economically but with adequate justification) an evaluation of the paper;
- the opportunity to see at close hand, a fruitful (one hopes) application of statistical methods to a real problem.

The last often generates a little surprise in the student, and then a growing enthusiasm, as it becomes apparent that expertise within his grasp can be useful in unexpected ways. The next to last benefit mentioned is the only one which relies on the writing of a report: in principle the carrying out of a project of this type without the requirement of a written report could be beneficial in all the other ways. However, the requirement to set down conclusions and arguments explicitly provides a discipline which reinforces the other requirements, and is valuable in its own right. Francis Bacon's dictum: 'Reading maketh a full man; conference a ready man; and writing an exact' applies powerfully here.

4.6.4 Statistical analysis

We take as an example of this kind of project the second part of the question about the 1981 French Presidential Election in Project 27 of Section 5.2; see Exhibit 9.

There is no single technique of analysis here which leads to a unique 'right answer' – indeed that is one of the virtues of this project: various approaches, at different levels of formality, can contribute something useful. We sketch some of these approaches below, without intending to be exhaustive. Work by a typical third-year student, or a student on a Master's degree programme, on the project could not be expected to cover fully all even of the following aspects – because of the demands that some of them make on time and technical background, for example – but a reasonably competent report ought to contain much of the rather informal discussion in (a)–(c) below and to include some analysis based on one of the more formal approaches (d)–(f). As a matter of interest we have seen versions

EXHIBIT 9

FRENCH PRESIDENTIAL ELECTIONS

In the 1981 French Presidential Election there were two stages: Round 1 on 26 April, and Round 2 on 10 May. Ten candidates stood in Round 1:

François Mitterand	A
Valéry Giscard-d'Estaing	B
Jacques Chirac	C
Georges Marchais	D
Brice Lalonde	E
Michel Crépeau	F
Arlette Laguiller	G
Michel Debré	H
Marie-France Garaud	J
Huguette Bouchardeau	K

In the second round there were only two candidates, namely A and B, who had been placed first and second in Round 1. Each received considerably more votes in Round 2 than in Round 1, the increases presumably coming from electors who had voted for other candidates or had abstained. It is of interest to know how A's and B's additional votes were made up from the possible sources.

The table [in Section 5.2, Project 27] shows the results of the election for 24 Departments of Metropolitan France, chosen systematically (every fourth Department) from an alphabetical list. All table entries represent thousands. The first column, headed 'Electeurs inscrits', gives the numbers of registered electors. The exact numbers of registered electors fell slightly between the two rounds but never by more than about 0.2%.

Report on the light you think the data throw on the question raised above, that is, where did the new votes which A and B gained between Rounds 1 and 2 come from?

---

of all the approaches suggested in (a) to (e) in students' work though not (yet) the multinomial approach of (f).

(a) Background information

It is useful to begin by finding out something of the background to the data: indeed a student report which did not make at least some attempt to put the data into a context would certainly be less than ideal, and some critical comment ought to be made. The point applies quite generally, of course, to almost any piece of statistical work. In the present case contemporary newspaper accounts reveal the following facts. The parties or political inclinations of the ten candidates were:

A	F. Mitterrand	Socialist
B	V. Giscard d'Estaing	Union pour la Démocratie (Centre-right)
C	J. Chirac	Rassemblement pour la République (Gaullist)

D	G. Marchais	Communist
E	B. Lalonde	Ecology Party
F	M. Crépeau	Left Radical Movement
G	A. Laguiller	Lutte Ouvrière
H	M. Debré	Independent Gaullist
J	M-F. Garaud	Independent Gaullist
K	H. Bouchardeau	Unified Socialist Party

In a simple left–right ordering therefore candidates Giscard d'Estaing, Chirac, Debré, and Garaud were right of centre, and Mitterand, Marchais, Crépeau, Laguiller, and Bouchardeau left of centre. Lalonde, of the Ecology or 'Green' movement, which had been gaining support in recent years in West Germany as well as France, is harder to place. Many of the minor candidates in these elections stand with the aim of influencing the policies of the leading contenders rather than with any serious hope of winning the Presidency themselves. As a price for advising their followers to vote for one or other of the second-round candidates they might be able to extract policy concessions or even the promise of cabinet seats. Six days after Round 1, for example, Marchais advised his followers to vote for Mitterand in the second round, having received assurances about the presence of Communist ministers in a future government of the Left. By 4 May all the left-of-centre candidates were reported as having endorsed Mitterand. Giscard d'Estaing, on the other hand, received only lukewarm support from the official Gaullists. Chirac announced on 1 May that he would personally be voting for Giscard, but advised his followers only that they should vote according to their consciences. A week later he was saying that it was his supporters' duty to stop Mitterand, but he appeared to have little enthusiasm for the alternative. The Green candidate Lalonde made no positive recommendation about whom his followers should support. Instead he gave a list of twelve questions on environmental issues and suggested that his followers' choice should be guided by the responses of the two candidates to them. Almost immediately Giscard's government announced a programme of ecological research. Opinion polls taken between Rounds 1 and 2 suggested that most Communists would support Mitterand, that not all Gaullists – perhaps only about three-quarters – would support Giscard, that the Ecology vote would split between the two candidates, and that a substantial proportion of voters – possibly one-third – who had abstained in Round 1 intended to vote in Round 2.

(b) Aggregate results

In the 24 sample Departments Giscard d'Estaing received more first-round votes than any other candidate – 23.06% of electors. Mitterand came second with 21.47%, followed by Chirac and Marchais. The complete results for the 24 Departments were (%):

Mitterand	21.47
Giscard d'Estaing	23.06
Chirac	13.88
Marchais	11.76
Lalonde	3.16
Crépeau	2.07
Laguiller	1.86
Debré	1.31
Garaud	1.08
Bouchardeau	0.92
Abstained or spoilt papers	19.43

These differ very little from the results for the whole electorate, even though, as shown in Fig. 4.1, the sample of 24 Departments is far from uniformly disposed over the area of Metropolitan France.

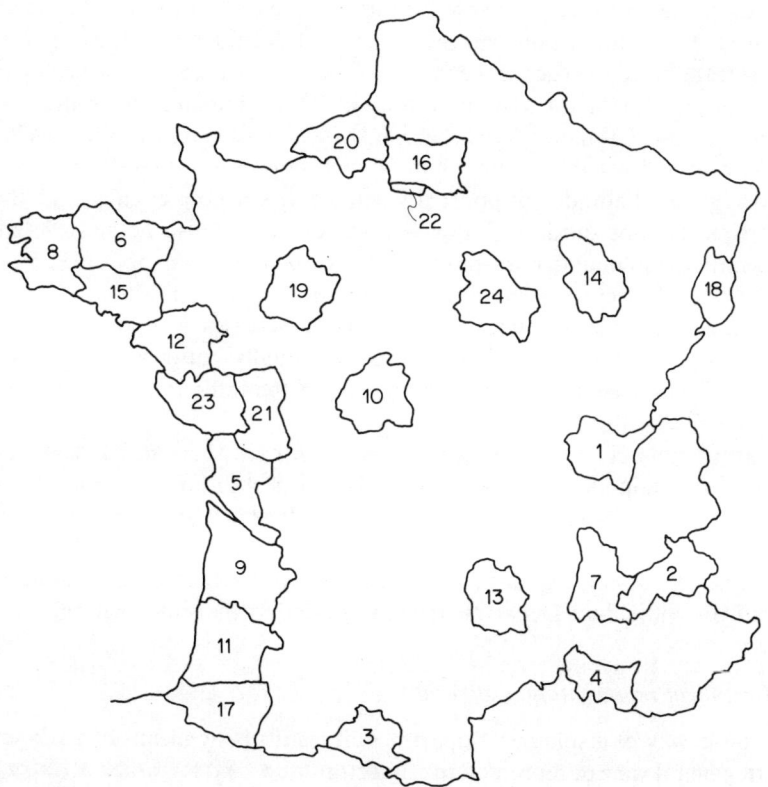

Figure 4.1: Locations of the 24 Departments within Metropolitan France.

In the second round the corresponding percentages were (ignoring the very small change in the electorate between the two rounds):

Mitterand	43.77
Giscard	40.43
Abstained or spoilt papers	15.80

The total support for candidates of the Left in Round 1 (Mitterand, Marchais, Crépeau, Laguiller, and Bouchardeau) amounts to 38.08%, and for candidates of the Right (Giscard, Chirac, Debré, and Garaud) to 39.33%. If all Left votes had gone to Mitterand and Right votes to Giscard, then Mitterand's victory would have to be explained by his having attracted many more of Lalonde's supporters and the first-round abstainers than Giscard – 5.69% as compared with 1.10%.

It is certainly possible that most of Lalonde's 3.16% support and a large proportion of those first-round abstainers who later voted in the second round (at least 3.63% of electors) transferred allegiance to Mitterand, so that his victory would be ascribable to the Greens and to a greater tendency among Left than Right supporters to save their vote for the decisive second round. On the other hand if, as the opinion polls suggested, only about three-quarters of Chirac's support transferred to Giscard then the two candidates must have picked up their extra support – 5.69% for Mitterand and 4.57% for Giscard – from first-round abstainers, from Lalonde's 3.16% and, in Mitterand's case, from the 3.47% of electors who had voted for Chirac but, on this supposition, abandoned Giscard. Supposing that Lalonde's support divided approximately equally and that a proportion, say one-third, of Chirac's disenchanted Gaullists actually voted for Mitterand (the remainder abstaining) we would find that about $3.63 + 0.67 \times 3.47 \simeq 6\%$ of electors who were first-round abstainers must have voted in Round 2, and that – to account for the observed increases in support for the two candidates – they must have divided roughly equally between Mitterand and Giscard. In this scenario the explanation for Mitterand's victory could be laid firmly at the Gaullists' door.

An approximately equal division of support for the two candidates by those first-round abstainers who subsequently voted is not implausible, and the total number of these, 6% of electors, or roughly one-third of first-round abstainers, is consistent with earlier opinion polls. But this is only one – albeit quite a plausible one – of many possible explanations of the data, and it is natural to see whether results from individual Departments can provide further information.

(c) Graphical representation of results

One simple way of displaying Departmental results is by means of a triangular plot. In general we can represent any 3-vector $(\alpha_1, \alpha_2, \alpha_3)$ for which $\alpha_1, \alpha_2, \alpha_3 \geqslant 0$ and $\alpha_1 + \alpha_2 + \alpha_3 = 1$ as the point in an equilateral triangle whose perpendicular distances from two specified sides are proportions α_1 and α_2 respectively of the altitude of the triangle. From the geometry of equilateral triangles the perpendicular distance to the third side is then automatically α_3 times the altitude.

Suppose then that we summarize the results for each Department by means of 3-vectors of proportions. There are many ways in which this might be done, but one natural and illuminating choice is to take the proportionate division of the electorate into 'Left', 'Right', and 'Other' in Round 1 and into 'support for Mitterand', 'support for Giscard', and 'Abstentions/Spoilt Papers' in Round 2. Each Department thus gives rise to two points. These may be plotted on the triangular graph and joined by a directed line to illustrate change in the Department between the two rounds. Figure 4.2 shows such a plot for the first 4 Departments: Fig. 4.3 is an enlargement of the central part of the same plot for all 24 Departments. Of course if simple ways of plotting and viewing 3- or higher-dimensional plots were available – as no doubt they will soon be as appropriate software is developed – then corresponding plots in higher-dimensional simplexes could be used instead of these triangular plots, thus allowing a less drastic condensation of the data.

The simple two-dimensional triangular plots can nevertheless bring out various features of the data. Victory for Mitterand in a particular Department

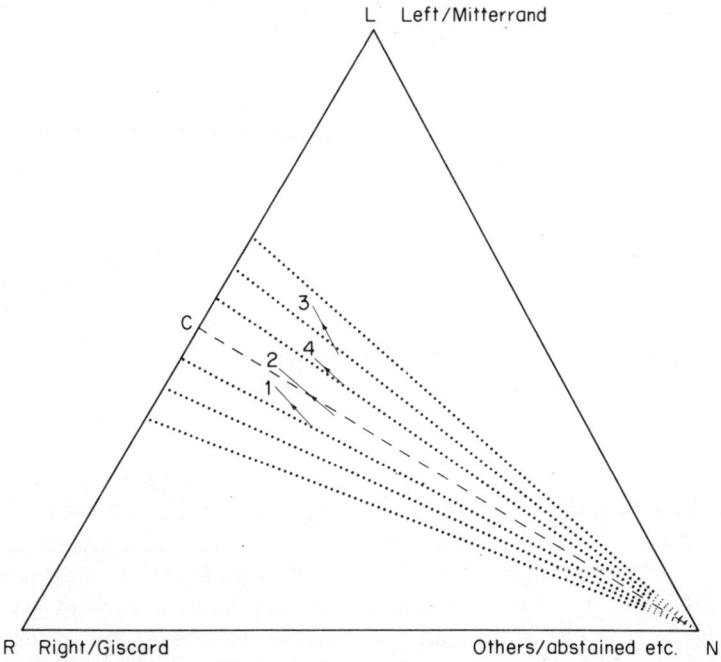

Figure 4.2: Triangular plot representing changes in voting patterns from round 1 to 2: Departments 1–4.

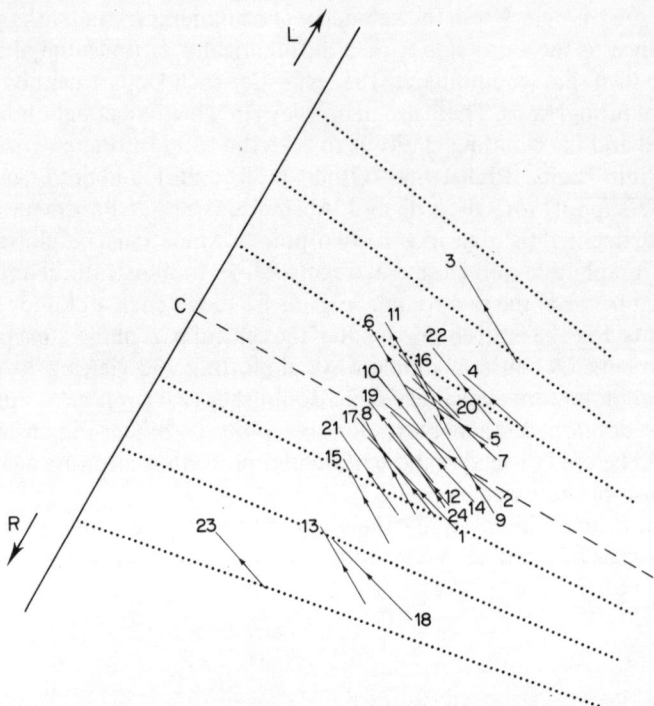

Figure 4.3: Central part of triangular plot representing changes in voting patterns from round 1 to round 2: all Departments.

corresponds, for example, to the second-round position of that Department lying above the bisector NC of the triangle. The distance of second-round points from the LR axis shows the proportion of electors who abstained in the second round. Since rays from vertex N in the figure correspond to lines of constant Left/Right division of support (as numbers of abstentions vary) the component of a Department's change in position in the direction at right angles to the ray from N through, say, its second-round position gives a measure of the 'swing' or change in Left/Right split between rounds. A reasonable approximation to this, convenient for a quick check, is the angle made by the Departmental line with the ray from N, since lengths of Departmental lines do not vary substantially. Figure 4.3 shows clearly that all Departments experienced a shift from Right to Left between the two rounds but that nevertheless Giscard came out the second-round winner in 9 out of the 24. The size of the shift was evidently variable. Some of the largest proportional shifts from Right to Left occurred in Departments 13 and 15 which had a Right majority large enough not to be completely eroded by large defections. The smallest shift was in the modestly left-wing Department 5.

If the opinion-poll suggestion that only about three-quarters of Gaullists would support Giscard in Round 2 were borne out in actuality then an apparent swing from Right to Left would be expected (unless there were a counteracting tendency for first-round abstainers and Green supporters to transfer differentially to the Right). Moreover, the movement might be expected to be most pronounced in those Departments with large first-round Right votes. Figure 4.3 does show some evidence of this feature, though the picture is by no means clear-cut, and there are obvious exceptions to the pattern (the left-wing Department 3, for example, which swung substantially further left in Round 2). Another way of looking at the feature is shown in Fig. 4.4, in which the Right-to-Left shift is plotted against proportional support for the official Gaullist candidate in Round 1. If Chirac's first-round supporters reflected their candidate's lukewarm attitude to Giscard we might expect there to be evidence of an increasing relationship between Leftwards shift and support for Chirac. The plot in Fig. 4.4 reveals that there is indeed such a tendency, and so lends credence to the opinion-poll findings.

It would be possible to pursue this line of argument in a quantitative way, taking into account also the change in numbers of abstentions, and finding numerical estimates for such quantities as the proportion of first-round Chirac supporters who failed to vote for Giscard, and the way in which those first-round abstainers who subsequently voted divided between Mitterand and Giscard. Such a quantitative analysis, however, is essentially the same as, and perhaps better viewed in terms of, a direct model for the transfer of votes, to which we turn next.

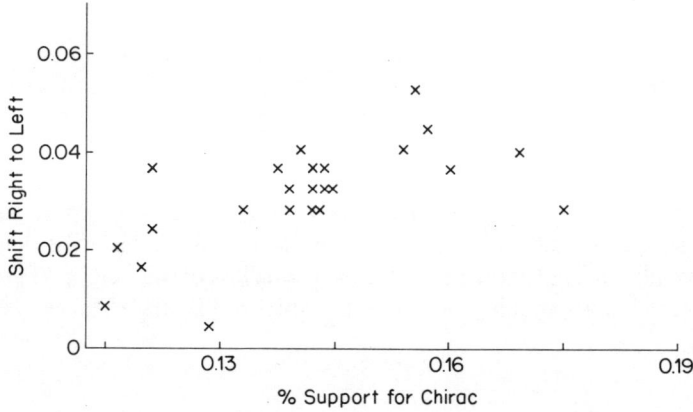

Figure 4.4: Right-to-Left swing vs. support for Chirac.

(d) Accounting equations and their estimation

Let us denote by X_{ik} the number of people voting for candidate i in the first round in Department k ($k = 1, \ldots, 24$). For convenience we think of abstentions and spoilt papers as votes for a fictitious 'Abstainers' candidate, and so i runs from 1 up to 11. If a proportion θ_{ijk} of those in Department k voting for i in the first round subsequently vote for j in the second round then the total number of votes for candidate j in the second round in Department k will be

$$Y_{jk} = \sum_{i=1}^{11} X_{ik} \theta_{ijk} \qquad \begin{matrix} j = 1, 2, 3 \\ k = 1, 2, \ldots, 24 \end{matrix} \qquad (1)$$

where Y_{3k} denotes number of abstentions. These are the so-called 'accounting equations'. The proportions θ_{ijk} are the objects of interest for answering questions about shifts of allegiance between the two rounds, but as they stand they are specific to each Department and are not estimable without more detailed data or further assumptions.

One way forward is to assume that the transition proportions θ_{ijk} vary from Department to Department around some country-wide average

$$a_{ij} = E(\theta_{ijk}) \qquad (2)$$

say. Then (1) may be written as

$$Y_{jk} = \sum_{i=1}^{11} X_{ik} a_{ij} + \varepsilon_{jk} \qquad (3)$$

for some 'errors' ε_{jk}, and the problem is to estimate the a_{ij} subject to the constraints

$$\begin{matrix} a_{ij} \geqslant 0 \\ \text{and} \quad \sum_{j=1}^{3} a_{ij} = 1 \quad \text{for each } i = 1, \ldots, 11, \end{matrix} \qquad (4)$$

which follow from (2) and the definition of the θ's.

If the constraints are ignored then it is natural to estimate the a_{ij} by least squares regression. Since error variances corresponding to larger Departments might be expected to be greater than those corresponding to smaller Departments, some form of weighting in the regression – in relation, say, to the total number of electors – seems appropriate. We might take, for example, variance proportional to electorate. Hawkes (1969) found that this approach gave worthwhile results for selected UK General Election data, despite its neglect of the constraints. In fact for the present data, regardless of the weighting used, this simple regression approach gives estimates which satisfy the *equality* constraint in (4), because $\sum_{1}^{3} Y_{jk} = \sum_{1}^{11} X_{ik}$ for each k, so that

$$E(Y_{3k}) = \sum_{1}^{11} X_{ik} a_{i3}$$

may be written alternatively as

$$E(Y_{3k}) = \sum_{1}^{11} X_{ik}(1 - a_{i1} - a_{i2}).$$

(If, however, the electorate had changed substantially between rounds – as it did in the data Hawkes analysed – then this simple relation would no longer hold.) On the other hand, there is no guarantee that estimates will satisfy the inequality in (4). To take account of the full set of constraints, both equalities and inequalities, the problem could be formulated as one in quadratic programming. Details of such a formulation (for UK election data, and with the assumption of constant error variance) are given by McCarthy and Ryan (1977).

Most students working on this project, however, are likely to try the cruder but simpler regression approach, at least initially. Typical results are shown below. The estimates are obtained from separate regressions of $\{Y_{1k}\}$ and $\{Y_{2k}\}$ on the $\{X_{ik}\}$, assuming that errors for different Departments are independent and have variances proportional to the total Departmental electorate, as suggested by Hawkes (1969).

						Estimated transition proportions a_{ij}					
	i	A	B	C	D	E	F	G	H	J	K	'Abstainers'
j	A	1.1	−0.1	0.2	0.9	−0.2	0.7	2.4	−0.5	0.7	1.7	0.2
	B	−0.1	1.1	0.8	0		0.3	0.2	−0.4	1.5	0.2	0.1	0.2

As a crude way of taking the inequality constraints into account a stepwise procedure suggests itself, in which parameters whose estimates lie outside the permitted [0, 1] range are set equal to 0 or 1 and the model refitted. This leads eventually (after several refits, adjusting one parameter – the most extreme – at a time) to:

						Re-estimated transition proportions a_{ij}					
	i	A	B	C	D	E	F	G	H	J	K	'Abstainers'
j	A	1.0	0.0	0.25	0.98	0.51	0.80	1.0	1.0	0.0	1.0	0.09
	B	0.0	1.0	0.77	0.0	0.50	0.25	0.0	1.0	0.0	0.0	0.17

Although some of these estimates are plausible and accord with the findings from other approaches, others – most glaringly that for candidate H – now violate other constraints. The stepwise procedure does not give standard errors of its estimates, but it is clear from this violation of the constraints that, for candidate H at least, their precision may be rather low. On general grounds, though, we might expect those transfer proportions which correspond to the major first-round figures to be estimated more precisely than those corresponding to candidates with lower first-round support. In fact the nominal standard errors of a_{31}, a_{32}, and a_{41} from the formal regression fitting were 0.06, 0.05, and 0.03 respectively and had stayed close to these low values throughout the stepwise estimation, suggesting that the likelihood surfaces are fairly sharply peaked in the a_{31}-, a_{32}- and a_{41}-directions, and that this peakedness is little affected by sectioning at different values of those parameters eventually set at 0 or 1. On these grounds then it seems fair to interpret the stepwise estimates as suggesting once again that although essentially all Communists backed Mitterand, a substantial proportion of Chirac supporters failed to vote for Giscard; these estimates imply, however, that all those not supporting Giscard voted for Mitterand, which seems perhaps less plausible than the earlier suggestion that many of them abstained.

(e) Inequalities for transfer proportions

Let us return to equation (1) and consider the information it contains about the transfer proportions $\{\theta_{ijk}\}$. We concentrate for the moment on transfers to Mitterand, $\{\theta_{i1k}\}$. For each department k the vector $\boldsymbol{\theta}_{1k} = (\theta_{i1k})$ belongs to the 11-dimensional hypercube $V = [0, 1]^{11}$ and the relevant equation (1) may be written as

$$Y_{1k} = \mathbf{X}'_k \boldsymbol{\theta}_{1k},$$

and interpreted as restricting $\boldsymbol{\theta}_{1k}$ to a 10-dimensional hyperplane in V. If there were a single set of transfer proportions $\boldsymbol{\theta}$ common to all Departments then the 24 hyperplanes generated by the different Departments would intersect at the point $\boldsymbol{\theta}$: if, as is undoubtedly the case, there is *no* common $\boldsymbol{\theta}$ but the $\boldsymbol{\theta}_{1k}$ for different Departments are dispersed around some countrywide average $\boldsymbol{\theta}$ then we might expect that the 24 hyperplanes will not intersect in a point but will be approximately coincident in some limited region of V containing $\boldsymbol{\theta}$. The inequality method described below is a simple and crude way of gaining information about this region by approximating to projections of it. The least-squares approaches discussed in the previous section also have a similar aim, but achieve it by different means.

We make the following assumptions: that those who supported Mitterand or Giscard in the first round maintain that support in the second round; that first-round supporters of other candidates of the Left either vote for Mitterand or abstain in Round 2 and correspondingly that first-round supporters of other candidates of the Right either vote for Giscard or abstain in Round 2; and that Ecology Party supporters and first-round abstainers may vote for either second-round candidate, or abstain. These are not the only possible assumptions, but they are not implausible and they will serve to illustrate the method. (In view of earlier findings it might be natural to allow the possibility that some Chirac supporters went further than merely abstaining in Round 2, and actually voted for Mitterand. A simple development of the approach below could be used to investigate this, but the calculations would be heavier.)

With the assumptions above, then, we have the following relation between first- and second-round votes for each Department (dropping the Departmental suffix for clarity)

$$Y_1 = X_1 + L(1 - \theta_{L3}) + X_5\theta_{51} + X_{11}\theta_{11,1}$$

where $L = X_4 + X_6 + X_7 + X_{10}$ represents the first-round votes for candidates of the Left other than Mitterand, and the θ's are the obvious transfer proportions. Since $0 \leqslant \theta_{51} \leqslant 1$ it follows that

$$Y_1 - L_1 - X_5 \leqslant X_{11}\theta_{11,1} - L\theta_{L3} \leqslant Y_1 - L_1 \qquad (5)$$

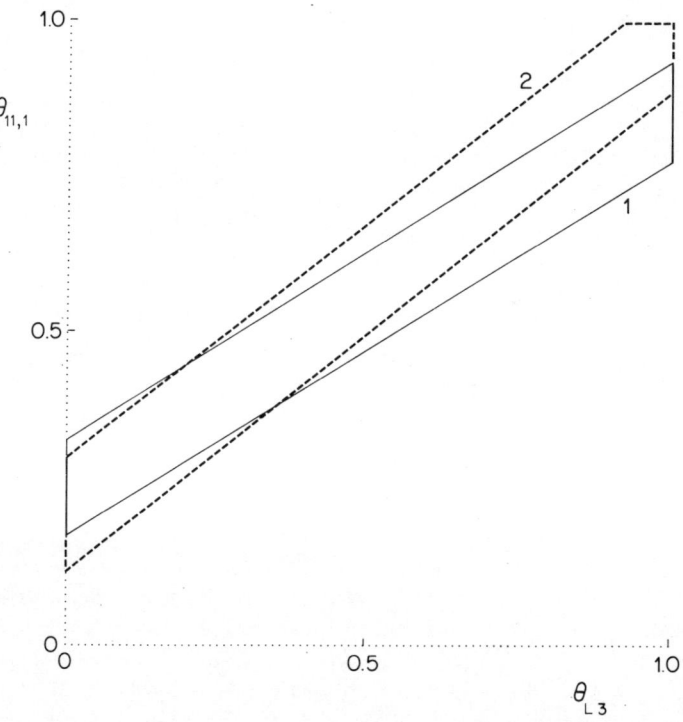

Figure 4.5: Regions to which transfer proportions $(\theta_{L3}, \theta_{11,1})$ are confined: Departments 1 and 2.

where $L_1 = L + X_1$. Thus feasible pairs of transfer proportions $(\theta_{11,1}, \theta_{L3})$ are constrained to lie in the region of the unit square defined by (5). Illustrations of these regions for Departments 1 and 2 are shown in Fig. 4.5. The intersection of the two regions of course consists of values of $(\theta_{L3}, \theta_{11,1})$ consistent with the outcome of the election in both Departments. By superimposing all 24 Departmental regions we can find which values of $(\theta_{L3}, \theta_{11,1})$ are most often consistent with Departmental results. It turns out that no value is consistent with results from all 24 Departments, or even with those from 23, but some values are consistent with 22 Departments (the exceptions being Departments 4 and 5). These values are those in the small wedge-shaped region in Fig. 4.6 bounded by $\theta_{L3} = 0$ and 0.06 and $\theta_{11,1} = 0.26$ and 0.32. Figure 4.6 shows the frequencies with which Departmental results are consistent with values of $(\theta_{L3}, \theta_{11,1})$ in the area bounding this region of maximum frequency. Similar calculations for transfers to Giscard lead to Fig. 4.7, showing frequencies for $(\theta_{R3}, \theta_{11,2})$ where R denotes the first-round votes for candidates of the Right other than Giscard. Here the values with which the greatest number of departments are consistent lie in the diamond-shaped region bounded by $\theta_{R3} = 0.16$ and 0.34 and $\theta_{11,2} = 0.15$ and 0.30. It is noteworthy that in the construction of Figs 4.6 and 4.7 a few Departments

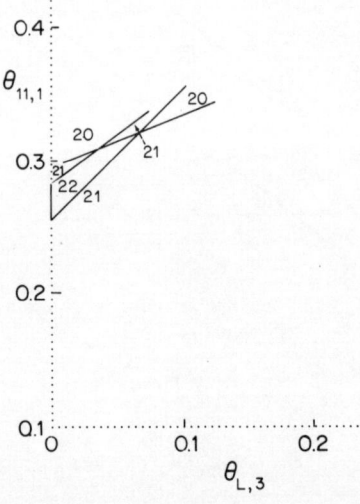

Figure 4.6: Frequency with which departmental voting was consistent with some values of transfer proportions $(\theta_{L3}, \theta_{11,1})$.

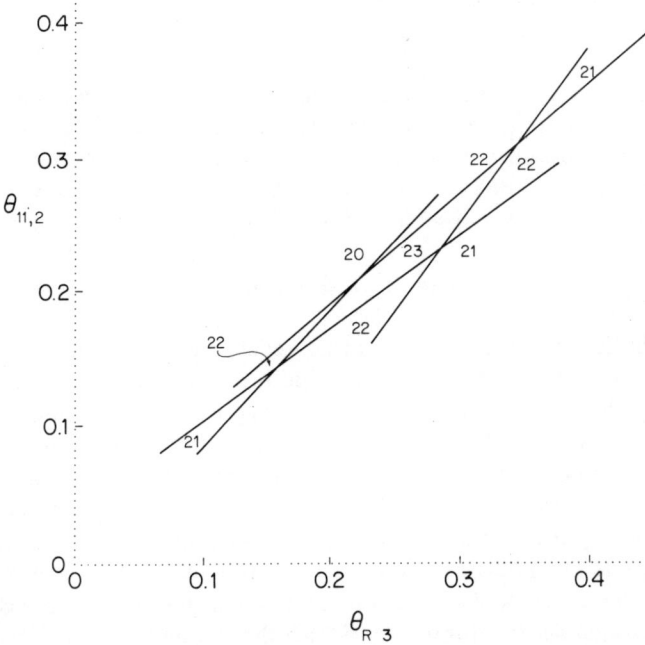

Figure 4.7: Frequency with which departmental voting was consistent with some values of transfer proportions $(\theta_{R3}, \theta_{11,2})$.

(notably Departments 4, 5, and 23) give regions which are disjoint or almost so from the rest.

Various conclusions might be drawn from this analysis: it might be seen as reinforcing the view that Left voters maintained their commitment more than did Right voters, that substantial numbers of first-round abstainers later came out to vote, and so on; but perhaps equally importantly it illustrates that there is apparently considerable variability between Departments in the transfer proportions, and in particular it raises the possibility that shifts in one or two Departments followed a somewhat different pattern from that common elsewhere.

(f) Multinomial-based approaches

The suggestion that follows goes back to Hawkes's paper on the analysis of electoral swing (Hawkes, 1969) referred to above. Let us write the number of first-round voters for candidate i in Department k as

$$X_{ik} = T_{i1k} + T_{i2k} + T_{i3k}, \qquad i = 1, \ldots, 11,$$

where T_{i1k} and T_{i2k} are the numbers voting for i in the first round and for A and for B respectively at the second round, and T_{i3k} is the number who initially voted for i and then abstained. Conditionally on X_{ik}, $(T_{i1k}, T_{i2k}, T_{i3k})$ is taken to be multinomially distributed with parameters X_{ik} and (a_{i1}, a_{i2}, a_{i3}), say. Though the T_{ijk}'s are not themselves observed, their totals

$$Y_{jk} = \sum_{i=1}^{11} T_{ijk} \qquad j = 1, 2, 3$$

are, and this leads to what is called an 'aggregate multinomial model' describing the probability structure of the data. The problem is now essentially one of estimating the two-way table of probabilities a_{ij} from observations of marginal totals. McCullagh and Nelder (1982, Section 8.6.2) discuss the use of a quasi-likelihood method on such problems; Hawkes had earlier advocated an approximate maximum likelihood method based on replacing the multinomial distributions by equivalent multivariate normal distributions.

As it stands the model assumes that Departments are homogeneous, in the sense that they all share the same probabilities a_{ij}, and we have seen earlier that this is questionable. A natural way to relax the assumption – by supposing that the $\{a_{ij}\}$ for different Departments have been drawn at random from some distribution of transition probabilities – has been studied in a London University MSc thesis by Firth (1982), and more recent work on such models, allowing also the possibility of dependence on covariates, is described in Brown and Payne (1986).

Epilogue

The foregoing suggestions cover a wide range of techniques and of sophistication. They are meant to indicate the breadth of analysis which this project-question can provoke, but in no sense are they meant to make up a contents checklist for a typical project report. The most satisfactory analyses from a purely technical point of view are undoubtedly those least-squares analyses which take account of all constraints, and the multinomial method and its generalization in the last section, but it would be unrealistic to expect most students to have the time to work through either of these approaches: indeed we have not worked through them either. In any case it is not certain that a technically highly sophisticated analysis is essential if worthwhile information is to be extracted from the data: as we have seen, quite naive methods can be productive, and most students make a useful analysis without going beyond them. The fact is that the important lessons to be learnt from projects such as this come as much from the student's experience of honestly confronting his existing technical competence with the demands of the data as from his acquiring new technical sophistication or routinely practising old. A report on this project which combined a relatively informal analysis, such as that in (a)–(c) above, with evidence of an appreciation – but not necessarily the carrying through – of some of the technically more sophisticated approaches could mark a very worthwhile advance in the student's practical competence.

CHAPTER 5

Projects

5.1 INTRODUCTION

In this chapter we give further examples of projects of the kind described and discussed in Chapter 4. Those given here illustrate further the variety that exists, but they are, and can be, no more than examples. There are not, of course, sufficient to cover all possible purposes, but even if there were it would be better for individual teachers to construct their own, particularly if students are not in British higher education: several of the projects are concerned with British data, such as those based on *Social Trends*, and should certainly, for those not living and working in Britain, be replaced by some with more direct relevance to the students' experience. Besides, although we have identified a number of types of project in Chapter 4, there are no doubt others that we have not thought of.

The projects are classified according to level in categories labelled 0 to 3, corresponding roughly to pre-university (0), first year (1), second year (2), third year and postgraduate (3), in the British system. Such a classification is at best only approximate, but it records, on at least an ordinal scale, our opinion of the level of sophistication (mostly of statistical knowledge, but to some extent of the general level of maturity and experience) required. Similarly, all projects are described as being of a certain duration – on the basis of one-sixth of a student's time being available – and of being appropriate for individuals or for groups. All of these classifications can, for most projects, be varied by explicitly allowing more or less time and by (explicitly or implicitly) requiring a more or less exhaustive treatment. There are also, for most projects, comments of diverse kinds. Aside from these comments, and the indication of level, incidentally, the projects are in the form in which they are given to our students. When used in class there is an oral introduction in most cases as well, but this will have a general purpose, and will not specify the task any more precisely. Baron and Worsdale (1980) provide a collection of assignments of somewhat similar type, though nearer to drill exercises, and it appears that they anticipate that students might be given an explicit set of tasks (which is mostly not the case in the projects here); e.g. calculate the expected breakeven quantity and expected monthly quantity.

Many of the projects here refer to data, and we have reproduced the data when it has not been published or it seemed likely that the source might not be widely available. The journal *Nature* is widely available round the world, for example, and we did not see the need to reproduce material from it (except for the detailed discussion of Project 23 in Section 4.6.3), but back numbers of *New Scientist* are probably not, and *Social Trends* is probably available in specialized libraries only. There are, incidentally, a few errors that we know of in the data, that are detectable from a study of the data alone, and these we have deliberately allowed to stand: anyone attempting the projects ought to find and comment on them. There is in fact a strong argument for inserting occasional (detectable) errors into data if they do not already exist, in order to keep students alert to the possibility that errors might occur, just as there is for making sure that there is, sometimes at least, some irrelevant information.

As we remarked at the beginning of this section, the projects given here are meant to be examples only, to be replaced, or at least supplemented, by others constructed by the individual teacher. This prospect raises two questions: what constitutes a project, and how are they to be found?

It is not, of course, possible to be very specific on these points, but one can certainly say something; in fact it is easier to say what is *not* a project. A project in the present sense has to be about a task in the real world, rather than something of a purely statistical or mathematical nature: the requirement to find the distribution of a binomial variable by simulation is not appropriate. Neither the problem nor the data need be strictly real, but they must be realistic. Unreal data which adequately mimic life are harder to construct than might be expected – one really ought, for example, to get orders of magnitude and relative proportions of figures in a company's budget about right. If the data or the context are too obviously unreal (widget manufacturing or testing of left-handed jam-jar openers) students, in our experience, will not treat them seriously enough, particularly at the feedback stage; even if such joke objects are not introduced, students are quite good at discovering artificiality and also lack of commitment – to the importance of the answers – on the part of the teacher. Again, the project should not amount merely to a problem which can be easily and uniquely translated into statistical language: 'the sex of different children in the same family may be assumed independent, but the probability of a male child is constant – find the effect of couples deciding to stop having children as soon as a boy is born'. This is not quite the same thing as saying that there should never be an 'obvious' approach to a project: it may, for example, be fairly obvious – to the teacher – that regression provides a sensible approach. But the question should not be posed that way, and a discussion of *why* regression is sensible should be expected in such a case, as well, of course, as translation of the results into a form which shows how and why the *original* (substantive) problem has been (partially) solved: although an unadorned histogram, for example, may tell a *statistician* all he needs to know, it would at the very least require some comment and interpretation before being useful in practice. For some time we experimented with problems which involved simulating a system, such as a car park or queues in a bank; but each time we found that the broader questions became submerged under the problems of programming, and valuable though some projects of this kind were in various ways, they were essentially developing techniques themselves rather than their use, and as far as the *present* goals were concerned we dropped them.

On the positive side, the subject matter and the answer to a good project will be of (potential) interest to someone who is not a theoretical or methodological statistician; it will in general require some thought as to how, and perhaps whether, statistics can help solve the problem, or, for projects which involve reviewing and summarizing other people's applications of statistics, whether the proposed methods are the best, or even appropriate; and it is almost certain to be obviously open-ended, in the sense that techniques and ideas to be introduced are not prescribed at the outset and, indeed, that there will be no way in which a limited discussion can finally and firmly resolve the substantive question.

A rather different approach to the appraisal of a potential project is to see which of the skills listed in Chapter 2 are likely to be fostered by it; it would be a miracle if all were, but if a reasonable proportion are (or can be by a suitable choice of the detailed arrangements) called into play, then the project may well be a useful one. Once the context in which any particular project will be used is known, it will usually be effectively obvious which skills are likely to be called into play. There are of course, additional requirements such as feasibility within the specified time, an absence of unintended ambiguity, and so on.

Finding (or, perhaps more exactly, constructing) projects is not as easy as one would like, but if one keeps oneself alert to the need they can often be suggested by articles in newspapers and periodicals. Standard statistical journals are rarely helpful in this respect, but general scientific publications – such as *Nature*, or *New Scientist*, or *Scientific American* – are useful, as are periodicals specializing in some other area such as geography, biology, medicine and so on. Rendall and Wolf (1983) are rather unusual in discussing British (official) sources in parallel with elementary statistical methods, and their integrative assignments could be used directly as projects at levels 1 or 2, as well as suggesting others. The collection of data by Andrews and Herzberg (1985) may also give inspiration.

Once an idea and any necessary data have been found, it remains to formulate in detail the specific problem to be posed. The wording of each project will inevitably depend on its own special context, but we suggest that one general aim in constructing an assignment should be, wherever possible, to word the question in such a way that it will be natural for the student to start work on it (without specific instructions to do so) with some reasonably straightforward task – a visit to the library to seek out background information, say, or the calculation of a few summary statistics – and to progress to more demanding aspects, where appropriate, only afterwards. This pattern not only bolsters confidence and gives a breathing space – valuable to both weak and strong students alike – in which to reflect on what is being asked, but it also encourages the habit, noted in Section

4.6.4 as being desirable quite generally in approaching statistical problems, of viewing any problem in context before launching into detailed or sophisticated analysis.

It is of course true, almost inevitably, that one will judge a suggested project to be less than ideal in some ways, but that seems no reason to regard it as unacceptable: here, as nearly everywhere else, the best is the enemy of the good.

5.2 SAMPLE PROJECTS

This section contains 36 examples of projects suitable for use with students at various levels. They are arranged in an order representing, roughly at least, increasing sophistication. A grouping by type is also given below.

Projects

1 Future populations	19 Counting lampposts
2 Some applications of statistics in computing	20 The local weather
	21 Wind, rain, and sun
3 Anti-aircraft fire in the Second World War	22 Naive forecasting
	23 Weather forecasting
4 The plight of the whales	24 Eyesight of Glasgow children
5 Petrol and car mileage	25 Maps and distances
6 Astronomical statistics	26 Extinction of surnames
7 Social questions	27 French presidential elections
8 Delinquency and family size	28 Digestive system of sheep
9 Olympic medals	29 Future numbers in schools and universities
10 Lives of great men and women	
11 Unemployment figures	30 Prison population
12 The ultimate mile	31 The demand for mathematics teachers
13 Catching up the men	
14 Starting a new job	32 Ball games
15 Short-term memory	33 The feeding behaviour of spiders
16 Reading habits of students	34 Expedition logistics
17 Mustard and cress	35 Recounts in elections
18 Reaction times	36 Oil exploration

Projects listed by type

Read and explain	No.	Level
Future populations	1	1/0
Some applications of statistics in computing	2	1/0
Anti-aircraft fire in the Second World War	3	1/0
The plight of the whales	4	1/0
Astronomical statistics	6	1

Critical review	No.	Level
Weather forecasting	23	2/3
Ball games	32	3
Feeding behaviour of spiders	33	3

Experiment/survey/collect data

Short-term memory	15	2/3
Reading habits of students	16	2/3
Mustard and cress	17	2/3
Reaction times	18	2/3
Counting lampposts	19	2/3
Maps and distances	25	1/2/3
Extinction of surnames	26	3

Data interpretation and/or presentation

Petrol and car mileage	5	1/0
Social questions	7	1/2/3
Delinquency and family size	8	1/2
Unemployment figures	11	2
The local weather, Part (i)	20	2/3
Wind, rain, and sun, Part (i)	21	3

Criticize, Criticize/Analyse

Olympic medals	9	2
The ultimate mile	12	2
Catching up the men	13	2/3

Investigative

Lives of great men and women	10	2
The local weather	20	2/3
French presidential elections	27	2/3
Eyesight of Glasgow children	24	3/2
Wind, rain, and sun	21	3
Naive forecasting	22	3
Digestive system of sheep	28	3

Seek data, synthesize

Lives of great men and women	10	2
Future numbers in schools and universities	29	3
Prison population	30	3
The demand for mathematics teachers	31	3

Model and/or synthesize	No.	Level
Digestive system of sheep	28	3
Expedition logistics	34	3
Recounts in elections	35	3
Oil exploration	36	3

Complete a report		
Starting a new job	14	2/3

How many people will there be in this country in 10 years time, in 20 years . . . ? Will there be a greater proportion of elderly people than now, a greater proportion of women, more people requiring university education . . . ?

Questions such as these are obviously important to anyone – including government, health authorities, industrial organizations, manufacturing industry – concerned with planning for the future. The key to them all is the estimation or projection of the future population, and its age structure.

The article referred to below describes the basis on which demographic projections are made. Read it and

- briefly outline the method in a page or two at most

[or • come to the next meeting prepared to explain the method].

Is the method sure to give the correct answer? What other factors might be taken into account in the hope of improving it? What does the author claim to be the main usefulness of the method?

KEYFITZ, N. (1978) How crowded will we become?, in *Statistics: A Guide to the Unknown*, 2nd edn. (ed. J. M. Tanur *et al.*), Holden-Day, 1978, pp. 355–367.

∗   ∗   ∗

Level: 1/0   *Duration:* 1/2 weeks   *Type:* Individual/group

Comment: Similar use to Project 4, 'Plight of whales'; Project 6, 'Astronomical statistics'; etc. Projects of this general type (see Projects 2 and 3 as well) are useful in, for example, illustrating the scope of statistics and stimulating thought about the way it is applied, without demanding too extensive a technical background. The requirement to report orally or in writing gives both a focus for the activity and also some early practice in the organization and communication of material. The question, along with other similar ones, might be used as a basis for short seminar presentations, in which case the written outline would not be needed, or purely as a written assignment, in which case the second of the alternative tasks is redundant. The choice of seminar presentations might be attractive if the class is large, but can be divided into subsets each jointly preparing a short presentation: see Section 3.12.

Applications of statistical and probabilistic ideas can play a central role in fully utilizing the computer's power. Read the article below and

- briefly outline, in a page or two at most,

[or ● come to the next meeting prepared to explain]

two applications of statistics in computing.

Can you suggest other problems which the methods described might be useful in tackling?

M. E. MULLER (1978) Information, simulation and production, in *Statistics: A Guide to the Unknown*, 2nd edn. (ed. J. M. Tanur *et al.*), Holden-Day, 1978, pp. 458–465.

*   *   *

Level: 1/0   *Duration:* 1/2 weeks   *Type:* Individual/group

Comment: Similar use to Project 1, 'Future populations', etc.

PROJECT 3   *Anti-aircraft Fire in the Second World War*

The beginnings of the subject of operations research go back to the Second World War, when groups of mathematicians, statisticians, and scientists were brought together to tackle various operational problems concerned with the running of the war.

One such group was concerned with ways of improving the effectiveness of anti-aircraft fire. Its work is described in the article below.

Read the article and

- briefly outline, in a page or two at most,

[or • come to the next meeting prepared to explain]

how the group broke down its task, and what the usefulness of that breakdown was. Which parts of their work would you describe as 'statistical' and which 'probabilistic'? In what ways would tasks of the kind they were faced with be easier nowadays?

PEARSON, E. S. (1978) Statistics and probability applied to problems of anti-aircraft fire in World War II, in *Statistics: A Guide to the Unknown*, 2nd edn. (ed. J. M. Tanur *et al.*), Holden-Day, 1978, pp. 474–482.

*　　　*　　　*

Level: 1/0   *Duration:* 1/2 weeks   *Type:* Individual/group

Comment: Similar use to Project 1, 'Future populations', etc.

We are often told that the whale is becoming a threatened species, its numbers having declined drastically over the past half-century. But how do we know how many whales there are? How do you count a wild population hidden in millions of cubic miles of sea?

Read the article below and

- briefly outline, in a page or two at most,

[or • come to the next meeting prepared to explain]

three methods for counting whales.

Point out any limitations you can see in the methods and try to suggest ways of getting round them.

To what extent do you think that similar methods could be used to estimate: (a) the stock of herring in the North Sea, and (b) the number of grasshoppers in a meadow?

CHAPMAN, D. G. (1978) The plight of the whales, in *Statistics: A Guide to the Unknown*, 2nd edn. (ed. J. M. Tanur *et al.*), Holden-Day, 1978, pp. 105–112.

∗   ∗   ∗

Level: 1/0   *Duration:* 1/2 weeks   *Type:* Individual/group

Comment: See Project 1, 'Future populations'. For this project in particular (though also to some extent for others of a similar type, such as Projects 1, 2, 3, and 6) the level is by no means limited to 0/1: the same question could be expected to exercise the same abilities (comprehension, imagination, criticism, presentation) in more advanced students, and, besides, might elicit a more technical response (independence, difficulties of random sampling, properties of estimators, etc.). With more advanced students a more generous page-allowance might be reasonable.

113

PROJECT 5   *Petrol and Car Mileage*

A family motorist bought a new car in August 1980 and, in his initial enthusiasm, decided he would try to keep a record of its petrol consumption and mileage. Accordingly, every time he bought petrol he entered the amount, the cost, and the car's current mileage in a small record book, a copy of which is shown in Table 1.

Two years later, while attending some evening classes on car maintenance and ownership, he casually mentioned his record book and was immediately asked by the lecturer and other members of the class to give a short (5–10 minute) talk on it two weeks later.

Suppose that he asks you, as a friend knowledgeable about statistics, for help in analysing and summarizing his data and in presenting them to an enthusiastic but not statistical audience.

Analyse the data in ways that you think would be of interest, and draft brief notes and graphical displays for the talk. Bear in mind that the audience might well be interested in such things as the rise in the price of petrol, changes in mileage with time of year, etc., as well as petrol consumption.

✳ ✳ ✳

Level: 1/0   *Duration:* 1/2 weeks   *Type:* Individual

Comments: Some further questions that might be of interest are:

- Does consumption vary greatly (on these data) with type of driving – for example, between town driving (longish intervals between fill-ups) and longer-distance driving (short intervals between fill-ups)?
- The Government test figure for petrol consumption in urban driving for this make of car was 38 mpg. How does this figure relate to actual experience?

The motorist's records shown below reflect among other things the gradual introduction of metric petrol pumps in the UK from early 1982.

114

Table 1 Fuel consumption data

Date	Gals*	Mileage	Cost (£)	Date	Gals*	Mileage	Cost (£)
6/ 8/80	8.60	23		18/ 7/81	4.10	5873	6.40
11/ 8/80	7.80			24/ 7/81	5.94	6067	9.67
13/ 8/80	7.44			28/ 7/81	6.00	6302	9.72
4/ 9/80	6.90	849	8.90	31/ 7/81	6.20	6541	10.30
17/ 9/80	6.61	1036	8.46	6/ 8/81	8.00	6848	13.11
30/ 9/80	6.00	1230	7.62	19/ 8/81	6.03	7017	10.15
12/10/80	7.93	1434	10.00	24/ 8/81	6.00	7287	10.10
26/10/80	6.09	1615	7.67	3/ 9/81	6.70	7517	11.20
8/11/80	5.98	1791	7.54	13/ 9/81	6.63	7755	11.00
12/11/80	4.00	1931	5.00	1/10/81	3.01	7976	5.00
16/11/80	8.07	2189	10.00	13/10/81	8.01	8079	13.21
22/11/80	4.88	2337	6.10	6/11/81	8.00	8305	12.96
7/12/80	6.64	2521	8.30	21/11/81	8.89	8516	14.81
21/12/80	6.61	2694	8.53	20/12/81	32.97l†	8712	11.50
15/ 1/81	8.00	2906	10.31	8/ 1/82	30.52l	8932	10.65
3/ 2/81	8.00	3104	10.66	19/ 2/82	37.92l	9329	12.25
21/ 2/81	7.60	3354	10.03	13/ 3/82	6.09	9481	9.20
2/ 3/81	7.00	3601	9.23	25/ 3/82	31.34l	9764	11.00
17/ 3/81	6.00	3739	8.87	9/ 4/82	39.42l	9986	13.60
31/ 3/81	8.00	3977	11.83	1/ 5/82	34.63l	10230	12.40
15/ 4/81	7.00	4214	10.50	21/ 5/82	22.70l	10359	7.96
22/ 5/81	7.00	4815	10.25	18/ 6/82	8.20	10609	13.89
26/ 5/81	2.60	4917	3.80	4/ 7/82	5.67	10791	9.75
6/ 6/81	5.58	5118	8.20	19/ 7/82	6.06	10999	10.00
16/ 6/81	6.00	5384	9.36				
25/ 6/81	6.00	5580	9.36				
12/ 7/81	8.05	5732	12.57				

* Imperial gallons
† l = litres

Look at the sky on a clear night and you see very large numbers of stars scattered apparently at random in every direction. For this reason, if for no other, it is hardly surprising that astronomers should have found statistical and probabilistic ideas useful in their studies.

The references below describe some applications of probability and statistics to astronomical questions. Read them and

● briefly outline

[or ● come to the next meeting prepared to explain]

how probability arguments were used to suggest: (a) the existence of double stars, (b) a common origin for the planets in our solar system, and (c) how disturbances in the sun's atmosphere can be studied by means of a correlation coefficient.

WHITNEY, C. A. (1978) Statistics, the sun, and the stars, in *Statistics: A Guide to the Unknown*, 2nd edn. (ed. J. M. Tanur *et al.*), Holden-Day, 1978, pp. 450–457.
PEARSON, K. (1978) Pierre Simon Laplace: 1749–1827, in *A History of Statistics in the 17th and 18th Centuries* (ed. E. S. Pearson), Griffin, London, 1978, pp. 668 and 716.

✳   ✳   ✳

Level: 1   *Duration:* 1/2 weeks   *Type:* Individual/group

Comment: Could be used as a group project in the same way as Project 1, 'Future populations'; Project 4, 'The plight of the whales'; etc. Technically slightly more demanding than these, since it requires understanding of some probability calculations.

Discuss each of the following statements critically in the light of the information in *Social Trends* No. 10 (1980). A reference is given in each case to a table or chart which is (presumably) directly relevant. Do not forget however, that there can be relevant information in other tables or charts – e.g. numbers of children in school (Tables 4.1, 4.2, 4.4 of *Social Trends*) are greatly affected by the age structure of the population (Chart 1.3): it would be sensible to glance at the list of contents on p. 3 and to consult *Social Trends* a second time when you have got the problem clear in your mind. Carry out any technical analysis that you think would be useful and mention any other collections and/or analyses of statistics that would throw light on the statements.

(a) Life is getting easier for the criminal in this country: he is less likely to be caught and found guilty and, even if he is, his sentence will be lighter (Table 13.1, Chart 13.9).

(b) Inequalities in the distribution of wealth have hardly changed, and something needs to be done about it (Table 6.30, Chart 6.4).

(c) Britain's railways are becoming safer and her roads more dangerous (Table 8.20).

*   *   *

Level: 1/2/3   *Duration:* 1/2 weeks   *Type:* Individual/group

Note: *Social Trends* is an annual publication of the UK Central Statistical Office, giving selected statistics, and to a lesser extent some commentary, on British and Northern Irish Society.
          For convenience, the tables referred to in (a), (b), and (c) are reproduced below, by permission of the Controller of HMSO.

Comment: Statement (a) is discussed at length in Section 4.6.1, so we add only some brief general remarks here. The statements above are slightly extended versions of assertions that might be found as headlines in newspapers. Projects of this general type exercise numerate common sense without making large demands on technical statistical knowledge – though they do require some general knowledge of society. They also provide a convenient way of introducing students to the more important published collections of data – for example, in the UK, when introducing this or a similar project one might well discuss briefly, in addition to *Social Trends*, publications such as *Economic Trends*, *Annual Abstracts*, and the *Monthly Digest* as well as mention indexes and guides to such published data. In this connection the book *Statistical Sources and Techniques* by F. J. Rendall and D. M. Wolf (1983) might be found to be a useful reference: it gives an elementary guide to (British) official statistics, particularly those concerned with economic matters, and to their use.

The idea for questions of the type above came from examination papers used in the UK Civil Service Commission's competition for entry to the Government Statistical Service: we gratefully acknowledge the inspiration. Some further similar questions are given below. Those in (h), (i), and (j) are in fact taken from an advertisement for the competition in November 1976, and are reproduced here by kind permission of the Controller of HMSO. The references in statements (d)–(f) below are to Tables in *Social Trends* No. 10 (1980), and those in (g)–(j) to *Social Trends* No. 3 (1972).

Newspaper headlines are a continuing source of further similar examples.

Further questions:

(d) The cost of educating a student at University is less than at a Polytechnic or Teacher Training College (Tables 4.11, 4.20).

(e) One in five children in England and Wales is likely to witness its parents' divorce before its sixteenth birthday (Tables 2.8, 2.13).

(f) Women are more likely to suffer from mental illness than men (Tables 8.13, 1.2).

(g) Marriage is getting more popular (Tables 1.2, 2.9, 2.10).

(h) Commuters into London are being forced to travel by car as British Rail provides less seating accommodation (1972, Tables 124, 125).

(i) The likelihood of a murderer being acquitted or not charged has risen dramatically over the last five years (1972, Tables 141, 142).

(j) Judged by class size and improved qualifications of teachers, the quality of public education is improving (1972, Tables 84, 85, Chart p. 124).

Table 13.1 Offences recorded by the police: people proceeded against, and people found guilty

England & Wales, Scotland, and Northern Ireland Thousands

	1951	1961	1966	1971	1976	1977	1978
Serious offences:							
England & Wales[1]	547	868	1,307	1,666	2,136	2,637	2,561
Scotland	83	109	148	181	265[2]	301	277
Northern Ireland	8	10	15	31	40	46	46
People proceeded against:							
England & Wales							
Serious offences	114	193	250	351	457	474	470
Less serious offences	626	1,014	1,269	1,445	1,753	1,688	1,623
Scotland							
Serious offences	26	30	38	40	41	42	44
Less serious offences	87	142	169	186	184	171	184
Northern Ireland							
Serious offences	3	4	4	5	5	6	5
Less serious offences	41	50	39	39	41	44	48
People found guilty and people							
whose charge was otherwise proven:							
England & Wales							
Serious offences:							
Magistrates Courts	115	151	208	282	359	372	367
Crown Court	18	31	24	40	56	57	57
Less serious offences	584	970	1,213	1,366	1,657	1,573	1,510
Scotland							
Serious offences	24	29	34	37	37	38	40
Less serious offences	76	130	157	175	171	159	171
Northern Ireland							
Serious offences	3	3	4	4	5	5	5
Less serious offences	39	48	36	35	39	41	44

[1] Up to 1976 inclusive excludes criminal damage value £20 or under.
[2] Because of changes in recording methods, the Scottish figures from 1975 onwards are not strictly comparable with previous years. The figure for 1976 comparable with previous years was 242.
Source: Criminal Statistics. *Home Office:* Criminal Statistics (Scotland), *Scottish Home and Health Department:* Ulster Year Book. *Northern Ireland Office*

Chart 13.9 People aged 17 or over sentenced for serious offences: by sentence and type of order

Great Britain

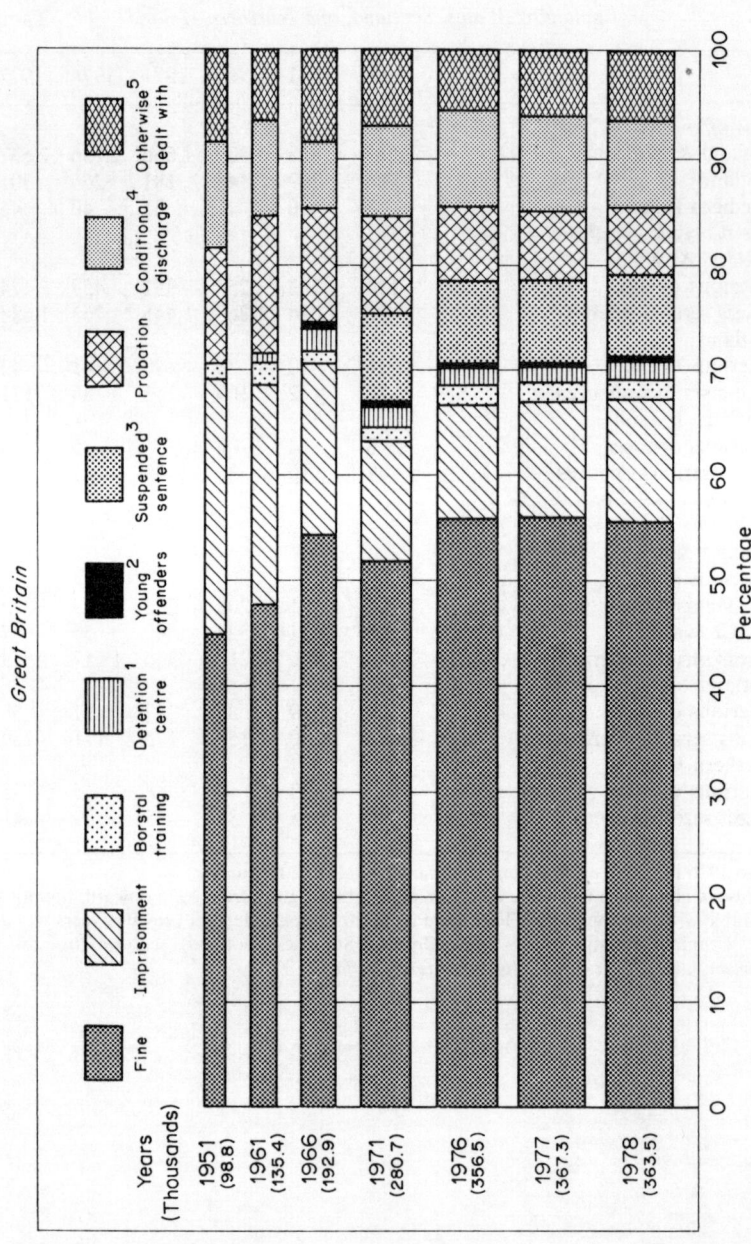

Footnotes on facing page

Chart 6.4 Dispersion of earnings[1]: men and women in manual and non-manual employment

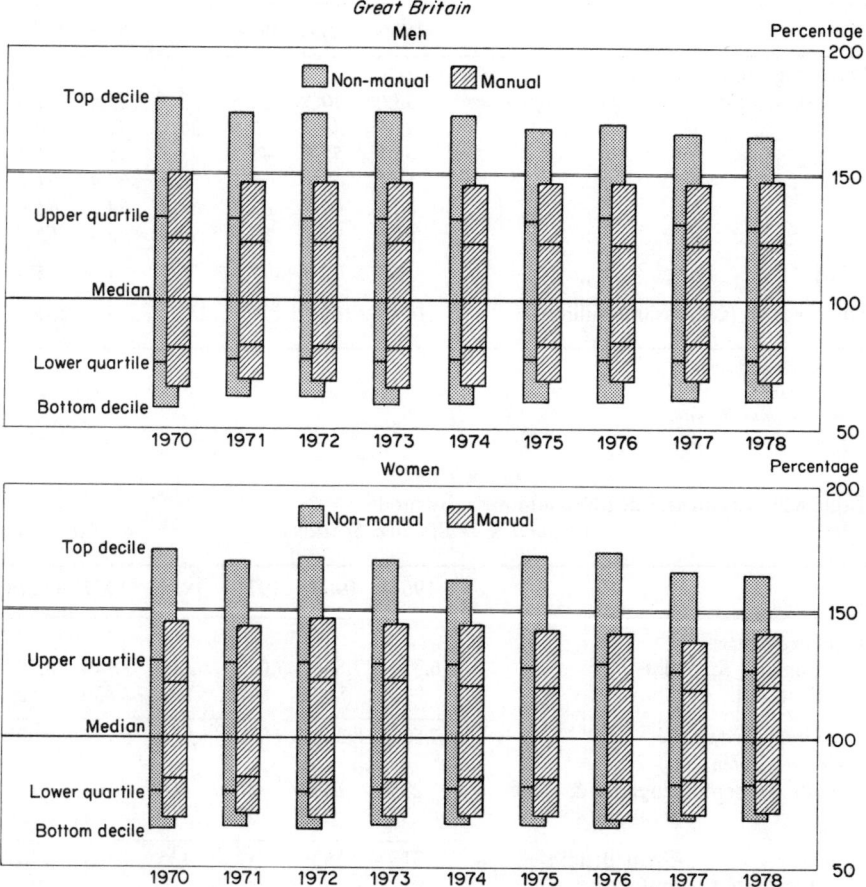

[1] Gross weekly earnings of men and women aged 18 and over in full-time employment whose pay for the survey period was not affected by absence. There are no data on a comparable basis for men for the years 1971 and 1972, so data are shown for men aged 21 and over.

Source: New Earnings Survey, *Department of Employment*

Footnotes to Chart 13.9

[1] For England and Wales only.

[2] Comprises for England and Wales – absolute discharge: for Scotland – absolute discharge and admonished.

[3] Detention centres were established in England and Wales in 1952 and in Scotland in 1960.

[4] Young Offenders Institutions were established in 1965 in Scotland. They do not exist in England and Wales.

[5] The suspended sentence came into effect in 1968 in England and Wales: it is not used in Scotland.

Source: Home Office; Scottish Home and Health Department

Table 6.30 Distribution of wealth

United Kingdom Percentages

	1966	1971	1974	1975	1976	1977
Percentage of wealth owned by:						
Most wealthy 1 per cent of population[1]	*33.0*	*30.5*	*22.5*	*23.5*	*23.7*	*24.0*
,, ,, 2 ,, ,, ,, ,,	*41.7*	*38.7*	*29.6*	*30.5*	*31.5*	*32.2*
,, ,, 5 ,, ,, ,, ,,	*55.7*	*51.8*	*43.1*	*43.8*	*46.1*	*46.4*
,, ,, 10 ,, ,, ,, ,,	*68.7*	*65.1*	*57.5*	*58.0*	*61.4*	*61.1*
,, ,, 25 ,, ,, ,, ,,	*86.9*	*86.5*	*83.6*	*83.3*	*84.2*	*83.9*
,, ,, 50 ,, ,, ,, ,,	*96.5*	*97.2*	*92.9*	*93.3*	*95.4*	*95.0*
Least ,, 50 ,, ,, ,, ,,	*3.5*	*2.8*	*7.1*	*6.7*	*4.6*	*5.0*
Total wealth (£ thousand million)[2]	107	164	236	272	296	345

[1] Aged 18 and over.
[2] End year.
Source: Inland Revenue

Table 8.20 Accidental deaths – summary: by mode

England & Wales, and Scotland Numbers

	1961	1966	1971	1976	1977	1978
Road accidents:						
England & Wales	6,778	7,546	7,072	6,115	5,943	6,772
Scotland	722	802	898	841	830	829
Great Britain	7,500	8,348	7,970	6,956	6,773	7,601
Railway accidents:						
All accidents: England & Wales	283	139	185	116	103	121
Scotland	30	23	27	19	21	26
Great Britain	313	162	212	135	124	147
Accidents to employees:						
England & Wales	140	55	50	37	32	28
Scotland	14	7	7	7	4	..
Great Britain	154	62	57	44	36	..
Home and residential accommodation accidents:						
England & Wales	6,882	7,470	6,245	5,619	5,369	5,341
Scotland[1]	1,262	1,113	800	788	895	722
Great Britain[1]	8,144	8,583	7,045	6,407	6,264	6,063
Accidents at work (non-transport accidents only):						
England & Wales	1,117	1,006	767	600	510	496

[1] Includes 'late effects' in Scotland.
Source: Office of Population Censuses and Surveys; General Register Office for Scotland

The following two letters appeared in successive issues of *The Times* in January 1973 and are reproduced by kind permission of their authors. Names have been changed.

From Mrs Beta: Sir, I was interested in the final point in Mr Alpha's letter [not reproduced here] entitled 'Unwanted children', where he wondered whether delinquent children more often came from large or small families. I sit as a magistrate on the Inner London Juvenile Court Panel and this same question occurred to me last summer.

Accordingly during my last quarter's sitting I have kept a tally of the number of children in the families of children in court whenever this information was given to the court. It is as follows:

Children in family	No. of cases
1	3
2	3
3	9
4	16
5	8
6	15
More than 6	16

I wonder whether juvenile court magistrates in other parts of the country have noticed a similar trend.

From Mr Gamma: Sir, your correspondent, Mrs Beta, who infers that delinquent children more often come from large families, would appear to have misread her own information. Does Mrs Beta not realize that, quite apart from any other factors, the average family of six children must stand six times more risk of raising a delinquent than a family with only one child? If the information that is given is recalculated to eliminate this loading, the table will appear as follows:

Children in family	No. of cases
1	3
2	1.5
3	3
4	4
5	1.6
6	2.5
More than 6	Less than 2.3

Draft a letter to *The Times* commenting on the statistical logic of Mrs Beta and Mr Gamma. You may wish to refer to information contained in Table 2.8 of *Social Trends* 1975.

Since such a letter must be relatively brief, set out your argument more fully in a longer report, incorporating any further analysis that you think appropriate.

$$* \quad * \quad *$$

Level: 1/2   *Duration:* 1/2 weeks   *Type:* Individual

Comment: The formal technical demands here are not great, but some maturity is perhaps required. A follow-up letter (from Mrs Beta) of the type asked for was in fact published in *The Times* on 11 January 1973: it is reproduced below with her kind permission. Students who find it without being told of its existence are, of course, to be congratulated.

From Mrs Beta: Sir, in replying to my letter of 2 January it is Mr Gamma and not I who has put a misleading interpretation on my own figures.

From the Registrar General's most recent available figures we find that large families are in the minority in this country (less than 10 per cent have four or more children) and that children from these larger families account for only approximately 22 per cent of the child population.

On my little chart they accounted for roughly 79 per cent of the children who appeared before my bench on remand cases (in which these details are noted as a matter of routine).

An article in *New Scientist* (23 August 1984, p. 30) discussed the distribution of medals in the 1984 (Los Angeles) Olympic Games, suggesting that gold, silver, and bronze should be roughly equal in numbers for a given country; a later letter in the same journal (13 September 1984, p. 59) suggested that the numbers need correcting for population size and Gross Domestic Product. The distribution of medals among the top 12 medal-winning countries is shown in the following table. How far are the arguments of both authors convincing? Can one conclude anything from these data?

	Distribution of medals			
	Gold	Silver	Bronze	Total
Canada	10	18	16	44
China	15	8	9	32
France	5	7	15	27
Great Britain	5	11	21	37
Italy	14	6	12	32
Japan	10	8	14	32
Netherlands	5	2	6	13
Romania	20	16	17	53
South Korea	6	6	7	19
United States	83	61	30	174
West Germany	17	19	23	59
Yugoslavia	7	4	7	18

*   *   *

Level: 2   *Duration:* 1 week   *Type:* Individual

There are a number of popular beliefs about the vital statistics of successful men and women: for example

(i) that births under various astrological signs are distributed differently for those eminent in different fields;
(ii) that eminent men and women cling to life determinedly in order to reach their birthdays;
(iii) that the likelihood of marriage (and/or divorce) depends on the fields of activity.

Investigate these beliefs and perhaps other similar ones.

The following sources of data might be useful:
Royal Society Obituaries 1932–54
Biographical Memoirs of Fellows of the Royal Society, Vol. 1 (1955)
Who Was Who
Dictionary of National Biography
Biographies of British Scientists
American Men and Women of Science

*   *   *

Level: 2   *Duration:* 4 weeks   *Type:* Group/individual

Comment: Further data relevant to (ii) on birthday and deathday is given in data set 71 of Andrews and Herzberg (1985). A related reference is the chapter by Phillips (1978) in *Statistics: A Guide to the Unknown*, 2nd edn. (ed. J. M. Tanur *et al.*), Holden-Day, 1978, pp. 71–85.

Suppose that you are a journalist on a (serious) newspaper and that your editor has come to the conclusion that trends in monthly unemployment figures are not well understood by the readers. Prepare an article, of not more than 1000 words, dealing with general questions such as whether to look at the change from last month to this (converted to an annual rate of change), or at the change in comparison with the same month a year ago, or at the change between the average level in the last three months and that for the same period a year ago. You should illustrate your article by the recent numbers of unemployed.

*   *   *

Level: 2   *Duration:* 1 week   *Type:* Individual

Comment: The suggestions made already recognize implicitly the fact that there are seasonal variations in unemployment, and, at least in part, cope with them. But any answer which did not make the problems of seasonal variation explicit would deserve criticism.

Other social or economic series, such as the Retail Price Index (Cost of Living Index), can be used in just the same way.

The trainer of a young, apparently world-class, mile-runner has just read the accompanying article [see Exhibit 10] by Trevor Kitson (*New Scientist*, 2 August 1984, p. 34), and has consulted you to find your opinion; the validity or otherwise of Kitson's analysis may affect his strategy over the next few years. What would your reply be?

✳   ✳   ✳

Level: 2   *Duration:* 1 week   *Type:* Individual

Comment: A letter to the editor from D. J. Finney expressing disagreement with Kitson's approach appeared in the issue of 16 August.

EXHIBIT 10

THE ULTIMATE MILE

A mathematical analysis of the record-breaking runs of the past suggests we may already be within one second of the fastest mile possible

Trevor Kitson

How fast is it possible to run the mile? Until that famous day in May 1954 at the Iffley Road track, Oxford, many people considered that 4 minutes would never be beaten. It was the unsurmountable barrier. Nowadays of course this barrier is broken so easily and regularly that it hardly rates a mention in the sports pages. Twenty-three years after Roger Bannister's effort, the 3 min 50 s barrier was first overcome (by New Zealand's John Walker). And now we appear to be fast approaching the next "milestone" of 3 min 45 s.

How long can such improvements continue? Not forever—there must surely be an ultimate time, less than which it is humanly impossible to run the mile. Perhaps it is possible to predict this time from a study of the physiology and anatomy of legs and lungs, the biochemistry of muscle action, the psychology of the "will to win", and so on, but the most obvious indicator of things to come is to look at things past.

I plotted the record since 1913 against the date, and immediately several interesting points become apparent (Figure 1). For instance, the long horizontal steps show when the record was ripe for a change—periods such as 1945–1954, just before the 4 min barrier disappeared, and 1967–1975, when Jim Ryun of the US held the record for so long. Similarly the large steps downwards show up some of the more impressive bites

---

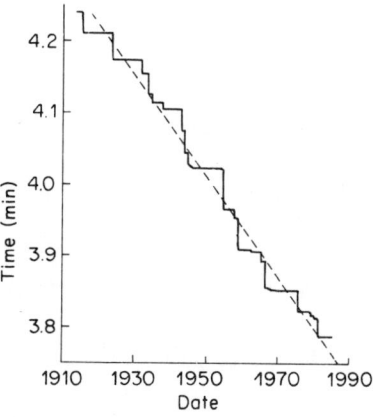

Figure 1

which were taken out of the record. The largest single reduction was 2.7 s by Herb Elliott of Australia in 1958.

However, the most striking observation is the remarkable linearity of the graph. Over the past 70 years the world mile record has been eroded at a surprisingly steady rate with little sign at first glance that this change is tailing off (but see below). The dashed line is the best straight-line fit to the data obtained by considering the value of the record as it stood on 1 January of each year from 1914 to 1984. This line (of correlation coefficient 0.991) may be represented by the equation:

$$t = 4.358 - 0.006933 \, T$$

(where t = the time for the mile in minutes, and T = the date minus 1900). It predicts that the 3 min 45 s mark will fall in 1987 and that the mile will be run in 3 min 30 s by 2023. Indeed, projecting the line all the way to the axis allows us to foretell that on 1 August 2528 the mile will finally be run in no time at all, a feat which will presumably ruin athletics as a spectator sport.

To return to seriousness, it is of course ludicrous to consider that the straight line of Figure 1 will continue indefinitely into the future. Sooner or later it must tail off to a limiting value. Close examination of the later half of Figure 1 suggests that there is indeed a suspicion of curvature. In order to bring this out more clearly, the data for the record as it stood on 1 January and 1 June of each year since the 4 min barrier was broken (which seems as good an arbitrary starting point as any) were subjected to a general curve-fitting computer program written by my colleague Mike Hardman. The results are shown in Figure 2, in which the dashed curve is the best fit to the data and represents the equation:

$$t = 4.777 - 0.02039 \, T + 0.0001040 \, T^2$$

The procedure of curve-fitting the results for the past 30 years, instead of the past 70, gives a very different prediction for the near future than discussed above. It suggests that Sebastian Coe's current world record of 3 min 47.33s is likely to stand for about the next 3 years, and that thereafter the fall of the record will be minimal. The trend since 1954, leads to the conclusion that the "ultimate mile" will be run in 1998 in a time of 3 min 46.66 s, only marginally quicker than today's record.

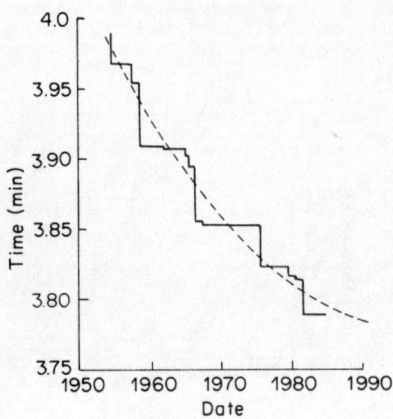

Figure 2

There are of course many imponderables in trying to predict the future. Will drugs ever become accepted and legitimate? Will there be further significant improvements in track surfaces and shoes? How will training methods change as progress is made in research into physiology, biochemistry, nutrition and so on? At present, though, the only hard data on which to make a prediction are the previous records, and the outcome depends on how long a time period is used as a basis for calculation. From Figure 1 it would be completely unremarkable if the 3 min 45 s barrier were smashed before 1990, but Figure 2 shows, perhaps rather surprisingly, that this mark may never be beaten and that we may already be within one second of the ultimate mile. Time alone will tell if the trailing-off since 1954 will continue or if the record will lurch downwards again to regain or undercut the straight line of Figure 1. I look forward to the next few athletics seasons.

Dr Trevor Kitson lectures in the Department of Chemistry, Biochemistry and Biophysics at Massey University, New Zealand.

It has been suggested (Kenneth Dyer, *New Scientist*, 2 August 1984, pp. 25–26) that the traditional views that women are prevented by physiological reasons from equalling men's performances in world-class sport is not necessarily correct, and that the gap (in, for example, world records) is continuing to close.

Investigate the situation as best you can.

Some data bearing on the question will be found in the *Encyclopaedia of Track and Field Athletics* (Mel Watman, 1981), but you may also be able to find other sources.

*   *   *

Level: 2/3   *Duration:* 2 weeks   *Type:* Group/individual

Comment: The fact that competitive women's sports has a shorter history than men's is relevant, and of course some events appeared in the women's calendar long before others; in any case, any inherent sex differential might well differ from event to event. A complicating factor is the relatively recent use of drugs which tend to make female athletes more masculine in physique and physiology. No answer to the question, clearly, can be definitive or even very convincing, and as in a number of other projects the discussion is more important than any detailed investigation. If this project is used for more senior students there would be the possibility of using rather sophisticated modelling techniques, but this should not be at the expense of elementary and general approaches.

You have just started work today in a new job as a statistician with an emerging young consulting firm. On your desk you find a small file of papers left by your predecessor before his hurried departure to take up a more lucrative appointment with a rival firm. Among these papers are parts of an uncompleted report which your new boss – in the hurried five-minute conversation you were able to snatch with him before he left on a 3-week business tour of the Middle East – has asked you to finish. 'It shouldn't take more than an hour or two', he had said. 'All the analysis has been done – it's just a matter of tidying the thing up, putting in some conclusions, and sending a completed draft report off to the client, who by the way is clamouring for results. Don't waste time on any more analysis, but if any ideas strike you about what's already been done, or about the way it's been written up, jot down a memo about it for the file – in case we get any come-back. By the way, if you can get this off tomorrow, there's a good chance you'll be asked to present the results at a meeting in Paris the week after next.'

The partially completed report, which is attached [see Exhibit 11], contains, you discover, an introduction, the data, and details of the analysis. Draft out further sections as your boss asked, and a separate memo containing any further comments for future use.

* * *

Level: 2/3   *Duration:* 2 weeks   *Type:* Individual

Comment: The use of projects of this type is discussed in Section 4.3.1. The overall aim is to give practice in the construction of reports, and so some preliminary guidance on report-writing – based, say, on suggestions such as those in the Appendix – is desirable.

The data in Tables 1 and 2 of the draft report are from a paper by K. D. Duncan (1966), and are reproduced by kind permission of the Editor of *Ergonomics* and of the author.

EXHIBIT 11

DRAFT PARTS OF A REPORT ENTITLED: INFANTRY
ACCLIMATIZATION FOR A HOT COUNTRY

INTRODUCTION

In May and June 1963, 108 infantrymen of the British Army took part in an experiment to evaluate an artificial acclimatization technique which it was hoped would improve their physical performance in a hot climate. Half of the group (the treatment group) undertook a fortnight's programme of physical exercises in an improvised hot chamber in the UK. The other 54 men performed the same exercises in a room of similar size at normal temperatures. Both groups were then flown to Aden where their performance under considerable heat stress was assessed during a 7-day exercise in the desert.

Part of the exercise consisted of daily marches, of 3 miles on the first day and 7 miles on subsequent days. The numbers of men who fell out (either in need of medical attention or because they were unwilling or unable to keep up with their section) are given in Table 1.

Table 1 Numbers falling out

Group		Day						
		1	2	3	4	5	6	7
Artificially acclimatized	n	49	48	47	45	42	40	38
	C	2	7	6	3	0	1	0
	C_M	1	7	6	3	0	1	0
Control	n	49	49	45	42	38	37	26
	C	5	23	12	11	3	11	0
	C_M	5	15	12	11	3	9	0

n = men beginning march
C = men falling out on the march (for any reason)
C_M = men falling out on the march (treated by medical teams)

The numbers beginning the march varied from day to day as a result of casualties who were not returned to the exercise.

Table 2 gives the mean marching times for the two groups.

The present report analyses these results with a view to assessing the benefits of the artificial acclimatization treatment. Three aspects of performance are studied:

(a) the speed with which the marches were completed day by day;
(b) the endurance of the men in the sense of their ability to undertake the marches on successive days;
(c) the seriousness of condition of these obliged to drop out.

Table 2 Mean marching times (minutes)

	Day						
Group	1*	2	3	4	5	6	7
Artificially acclimatized	67	163	180	209	133	131	110
Control	75	179	160	194	143	149	110
S.e. of difference	4.21	7.90	12.65	16.29	6.52	6.69	3.85

* Day 1 was of 3 miles instead of the 7 miles on all other days.

Statistical methods used are:

for (a), simple normal significance tests;
for (b), χ^2 tests for frequency data and approximate normal confidence intervals;
for (c), informal inspection of percentages, and normal tests applied to simple scores.

ANALYSIS

(a) Times

Observed differences in mean marching times and approximate p-values for a one-sided test of a zero difference in their expectations are given in Table 3.

Table 3 Differences in marching times

	Day						
Group	1	2	3	4	5	6	7
Difference Artificially acclimatized − control	−8	−16	20	15	−10	−18	0
Approximate p-value (%)	2.8	2.1	94.3	82.4	6.2	0.4	50.0

The test is based on the assumption that under the hypothesis of zero difference in expectations the difference/s.e. follows approximately a standard normal distribution.

(b) Endurance

From the data on numbers beginning each day's march the frequencies for length of participation in the exercise are calculated, as in Table 4.

Table 4 Length of participation

		Days						
	1	2	3	4	5	6	$\geqslant 7$	Totals
Treated	1	1	2	3	2	2	38	49
Control	0	4	3	4	1	11	26	49

(i) Tests for homogeneity

Under the assumption that the distribution of endurance is the same for the two groups the expected numbers are as shown in Table 5.

Table 5 Expected length of participation

			Days			
1	2	3	4	5	6	$\geqslant 7$
0.5	2.5	2.5	3.5	1.5	6.5	32.0

We use the generalized likelihood ratio test statistic $L = 2\sum_{i,j} n_{ij} \log(n_{ij}/e_{ij})$ to test for homogeneity, where n_{ij} and e_{ij} denote observed and expected numbers in the (i, j)th cell of Table 4. If the two distributions are the same, L has approximately a χ_6^2 distribution.

$$L = 13.12 \quad \text{so} \quad p < 0.05.$$

The χ^2 approximation might be improved by pooling classes with low expectation, as in Table 6.

Table 6 Length of participation (pooled classes)

	Days			
	$\leqslant 3$	4 or 5	6	$\geqslant 7$
Treated	4	5	2	38
Control	7	5	11	26
Expected	5.5	5.0	6.5	32.0

This approach gives $L = 9.95$, which, on comparison with χ_3^2 still yields $p < 0.05$. Pearson residuals $(n_{ij} - e_{ij})/e_{ij}^{1/2}$ are shown in Table 7.

Table 7 Pearson residuals for Table 6

-0.64	0.00	-1.76	1.06
0.64	0.00	1.76	-1.06

To identify the source of the inhomogeneity more precisely consider the numbers surviving the whole exercise, as shown in Table 8.

Table 8 Numbers who completed and did not complete the exercise

	Endurance (days)		
	$\leqslant 6$	$\geqslant 7$	
Treated	11	38	49
Control	23	26	49

For this, $L = 6.6$, $p \simeq 0.01$. Evidently a larger proportion (78%) of the treated group than of the control group (53%) were able to complete the exercise.

The endurance of those who did not complete the exercise is shown in Table 9.

Table 9 Endurance in those who did not complete the exercise

	Days		
	$\leqslant 3$	4 or 5	6
Treated	4	5	2	11
Control	7	5	11	23

For this, $L = 3.3$, $p \simeq 0.20$. There is therefore no very strong evidence of a difference in endurance for those obliged to drop out before the end of the exercise. The large number (11) in the control group, however, who left after 6 days is noteworthy, although not significant. (Note that these two tests are independent and the values of L additive.)

(ii) Comparison of mean endurance of those who did not complete the 7 days

	Mean endurance	Sample variance
Treated (11)	3.9	2.57
Control (23)	4.5	2.83

Pooled variance $= 2.75$
Approximate 95% confidence interval for the difference
control $-$ treated $= 1.5 \pm 2 \times 2.75^{1/2}(1/11 + 1/23)^{1/2}$
$\qquad\qquad\qquad = 1.5 \pm 1.2$.

(c) Condition of those dropping out of marches

(i) Percentages

If we classify the condition of those dropping out as severe or mild according as medical attention was or was not received, then numbers are as shown in Table 10.

Table 10 Condition of men who dropped out

		\multicolumn{7}{c}{Day}						
		1	2	3	4	5	6	7
	Mild	1	0	0	0	0	0	0
Treated	Severe	1	7	6	3	0	1	0
	Total	49	48	47	45	42	40	38
	Mild	0	8	0	0	0	2	0
Control	Severe	5	15	12	11	3	9	0
	Total	49	49	45	42	38	37	26

Corresponding percentages are given in Table 11.

Table 11 Percentages for Table 10

		\multicolumn{7}{c}{Day}						
		1	2	3	4	5	6	7
Treated	Mild	2.0	0.0	0.0	0.0	0.0	0.0	0.0
	Severe	2.0	14.6	12.8	6.7	0.0	2.5	0.0
Control	Mild	0.0	16.3	0.0	0.0	0.0	5.4	0.0
	Severe	10.2	30.6	26.7	26.2	7.9	24.3	0.0

Thus on every day except the last a higher proportion of the control group dropped out, and a higher proportion received medical attention.

(ii) Seriousness of condition

If seriousness of condition of those who dropped out is scored numerically, say as

10 for a mild drop-out
20 for a severe drop-out,

then the average score can be taken as an 'index of suffering' (Table 12).

Table 12   Average index of suffering

	Day						
	1	2	3	4	5	6	7
Treated	0.6 (9.8)	2.9 (49.9)	2.6 (44.3)	1.3 (25.0)	0.0 (0.0)	0.5 (10.0)	0.0 (0.0)
Control	2.0 (36.6)	7.8 (63.4)	5.3 (78.6)	5.2 (77.7)	1.6 (29.0)	5.4 (73.5)	0.0 (0.0)

Figures in brackets are sample variances.

Rough (one-sided) tests of significance of the differences control − treated give the p-values in Table 13.

Table 13 p-values for comparisons in Table 12

Day	1	2	3	4	5	6
p	0.070	< 0.001	0.053	< 0.01	0.03	< 0.001

(iii) Failures to appear next day

The foregoing analysis takes account only indirectly of suffering serious enough to lead to complete withdrawal from the exercise. If it is assumed that men who failed to appear on the following day were among those who received medical attention (when any was given) then the numbers can be inferred to be as in Table 14.

Table 14   Conditions of men who dropped out, taking account of withdrawal from exercise

		Day					
		1	2	3	4	5	6
Treated	Mild	1	0	0	0	0	0
	Severe	0	6	4	0	0	0
	Did not appear next day	1	1	2	3	2	2
Control	Mild	0	8	0	0	0	0
	Severe	5	11	9	7	2	0
	Did not appear next day	0	4	3	4	1	11

With a score of 30 for failure to appear next day these give indices of suffering as in Table 15.

Table 15 Indices of suffering from Table 14

	Day					
	1	2	3	4	5	6
Treated	0.8 (19.8)	3.1 (59.1)	3.0 (63.3)	2.0 (56.0)	1.4 (40.9)	1.5 (42.8)
Control	2.0 (36.8)	8.6 (105.6)	6.0 (104.0)	6.2 (113.9)	2.4 (39.0)	12.7 (219.5)

Approximate p-values in a comparison of control with treated are shown in Table 16.

Table 16 p-values for comparisons in Table 15

Day	1	2	3	4	5	6
p	0.13	<0.01	0.06	0.02	0.24	$\ll0.001$

*   *   *

In this project you are asked to design, carry out, and analyse an experiment into factors affecting short-term memory.

A suggested form for the experiment is as follows. Subjects are given a list of unconnected words and allowed to study them for a fixed length of time. The list is then withdrawn and after another fixed length of time the subject is asked to repeat as many words from the list as he can. In designing the experiment there are many practical details you will have to consider: selection of subjects, randomization, lengths of study and retention times, words to be used, and length of list, etc. It is suggested that you use three factors each at two levels:

study time – short and long
retention time (time interval between list being removed and subject being asked to repeat words) – short and long
word list – well-known words and unusual words

You may find it helpful to carry out a brief pilot experiment to gain some knowledge on suitable words and times.

You may modify this experiment in any way you think would be of interest but you should discuss and gain approval for this first.

* * *

Level: 2/3   *Duration:* 4 weeks   *Type:* Group

Comment: As with most experiments or surveys this could be adapted for different levels by, for example, simplifying or making more complicated the number and structure of the factors studied. In the present example both the number and the levels of factors are prescribed and the project may for this reason be appropriate for students at an earlier stage in their training than some of the following examples.

(This project is based on an idea of Dr A. W. Bowman, to whom we are most grateful.)

Design and carry out a survey to discover the newspaper and periodical reading habits of students, and write a report describing your conclusions.

*   *   *

Level: 2/3   *Duration:* 4 weeks   *Type:* Group

Comment: This can be done at various levels of complexity, from trying to discover how many read a newspaper (but what is a newspaper? and what is to 'read a newspaper'? and is it about reading one every day, or occasionally?) to the investigation of which newspapers are read and even which sections are read first.

Mustard and Cress

At a later stage you will be asked to carry out an experiment concerning the growing of mustard and cress in the home, in order to investigate the effect on growth of such factors as the growing medium used, the amounts of light and nutrient provided, competition with the other crop, and any other conditions that you think might be of interest. Seeds, seed trays, plant nutrient, and various types of growing medium will be provided.

At this stage in the project you are asked to draw up plans for the experiment, deciding firstly on the particular aspect(s) of growth which you wish to investigate and secondly on the various treatments you will use. These plans will need to include, among other things, details of the experimental design that will be used, the conduct of the experiment, and the precise requirements for equipment and materials.

✳   ✳   ✳

Level: 2/3   *Duration:* 4 weeks   *Type:* Group

Comment: The aims are deliberately left rather vague: but, mustard and cress being a 'crop' used in salads and such like, the response variable(s) should reflect edibility considerations. For a fuller discussion of this experiment see Section 4.6.2.

At a later stage you will be asked to carry out an experiment on the length of time people take to react to a visual stimulus. The aim will be to investigate how reaction times are affected by such factors as the sex and age of the subject, the time of day, practice, distractions and any other conditions that you think might be interesting and relevant. A convenient and simple way of measuring the time is to drop a metre rule vertically through the subject's hand and to note where he succeeds in grasping it.

Initially you are asked to plan the experiment, deciding firstly on which factors and treatments you wish to investigate, secondly on their arrangement in an appropriate experimental design, and thirdly on the details of how the experiment will be carried out.

Some preliminary decision will be necessary at the next class meeting, with a view to making final decisions at the meeting after that.

*   *   *

Level: 2/3   *Duration:* 3 weeks   *Type:* Group

Comment: An alternative method for measuring reaction times, for those who
             have the equipment, is to use a microcomputer with an internal
             clock; the necessary randomness in the time at which the stimulus
             appears could also be provided this way, and the records collected
             and processed directly.

Counting Lampposts

Define a suitable urban area accessible from your place of study and no less than 2 kilometres square, and estimate as accurately as you can the total number of roadside lampposts within it. Report on the method you adopt and its result, and discuss in particular the accuracy of your estimate.

*   *   *

Level: 2/3   *Duration:* 4 weeks   *Type:* Group

Comment: The most obvious approach is to sample the network of roads in the area. In this case some advanced planning will be needed – to decide, in the light of the time and manpower available, the type and extent of sampling, and the organization of the fieldwork to be carried out. Stratification might, of course, be worthwhile, and in this connection it would be useful to establish whether there are different standards for the spacing of lampposts on different classes of road.

An attractive alternative approach would be to base an estimate on the amount of electricity consumed for lighting; but unfortunately, so far as we know, the information necessary for this is not usually available.

A further intriguing possibility that we have heard suggested is to try to apply capture–recapture methods to the problem.

The problem itself may appear somewhat eccentric, but is in fact part of a real one. In the late 1970s the then South Yorkshire County Council resolved to employ three people to count the lampposts in its area. Though the motivation was partly job-creation, there was also a wish to obtain information about street lighting (such as the location and type of lamps as well as their total number) that the Council did not at that time possess, records having never been centrally maintained since the introduction of municipal street lighting in the nineteenth century.

The data in Table 1 are extracts from the records of the weather station in Weston Park Sheffield for the period between 1963 and 1978.

(i) Summarize the main aspects of the local weather in ways that you think would interest the intelligent but non-statistical inhabitants of Sheffield.
(ii) Discuss whether there is any evidence of changes in the local climate. (It is possible, for example, that rainfall might have decreased and hours of sunshine increased as Clean Air policies took effect during the 1960s and 1970s.)

*   *   *

Level: 2/3   *Duration:* 2 weeks   *Type:* Individual

Comment: This is a project in which it would be desirable, if possible, to replace the data given here with a similar set from the students' own locality. Part (ii), of course, would probably then need modifications too. Clean Air legislation was passed in 1956 in the UK and smoke-free areas introduced gradually thereafter.

Table 1 Monthly temperature, rainfall and hours of sunshine at Weston Park Sheffield 1963–78
(Reproduced by kind permission of Sheffield City Museums)

Month	Temperature (deg. F)			Rainfall		Sun
	Max.	Min.	Mean	Inches	Days	Hrs.
1963						
Jan	43.7	15.8	30.5	1.83	14	39.4
Feb	42.9	18.0	30.3	1.11	11	41.9
Mar	57.8	20.0	41.5	2.84	17	93.8
Apr	63.3	32.4	46.4	2.37	18	46.4
May	75.2	37.0	50.9	1.24	16	155.1
Jun	78.0	43.7	58.7	3.90	17	179.5

145

Table 1 (continued)

Month	Temperature (deg. F)			Rainfall		Sun
	Max.	Min.	Mean	Inches	Days	Hrs.
1963						
Jul	80.7	51.1	59.6	2.70	10	173.3
Aug	74.8	45.0	57.5	2.36	18	88.5
Sep	76.1	39.1	55.7	3.12	11	136.6
Oct	67.1	38.1	51.0	1.55	12	73.4
Nov	54.9	31.7	45.5	5.03	25	35.3
Dec	49.0	20.0	37.6	0.73	13	35.1
1964						
Jan	52.1	22.0	38.1	0.91	8	54.8
Feb	54.9	22.3	39.9	1.13	12	50.5
Mar	50.5	26.9	37.9	4.81	20	46.6
Apr	67.8	32.0	48.3	1.81	16	91.3
May	74.8	41.1	56.1	2.18	13	203.0
Jun	74.1	39.0	56.4	2.76	18	105.8
Jul	75.6	46.4	61.8	2.00	12	184.9
Aug	80.7	41.5	60.0	1.94	10	179.3
Sep	73.4	42.2	57.5	0.82	8	165.5
Oct	62.8	32.3	48.3	1.64	13	106.7
Nov	58.0	27.2	44.9	1.70	15	40.9
Dec	58.0	23.9	38.0	3.69	18	61.2
1965						
Jan	51.5	29.5	38.5	3.55	22	57.4
Feb	48.7	25.9	38.4	0.87	16	21.3
Mar	74.4	17.4	42.0	2.91	12	112.0
Apr	68.0	34.0	46.9	1.97	17	134.7
May	79.8	33.4	52.5	2.69	15	137.1
Jun	73.2	43.0	58.6	3.26	16	163.7
Jul	70.3	43.5	56.8	2.10	15	91.3
Aug	74.8	44.0	59.0	1.93	14	133.7
Sep	71.1	40.9	54.4	6.59	18	82.8
Oct	70.1	37.0	51.6	1.08	9	73.1
Nov	54.9	26.1	39.6	5.80	18	63.1
Dec	55.5	25.5	40.1	8.26	22	52.1
1966						
Jan	53.0	20.1	36.5	1.89	17	22.9
Feb	56.7	28.2	40.6	6.07	22	25.2
Mar	54.0	30.2	44.0	1.84	15	109.3
Apr	71.0	31.0	42.0	5.53	22	74.1
May	77.3	40.3	52.6	2.31	11	209.2
Jun	75.1	47.2	59.9	2.41	12	142.8
Jul	74.8	46.8	59.0	1.75	14	130.8

Table 1 (continued)

Month	Temperature (deg. F)			Rainfall		Sun
	Max.	Min.	Mean	Inches	Days	Hrs.
1966						
Aug	78.1	43.0	58.7	3.66	15	124.6
Sep	72.2	41.0	57.3	2.19	9	107.8
Oct	69.9	34.3	50.0	3.99	17	66.1
Nov	52.7	32.6	41.7	3.70	19	41.1
Dec	57.0	27.7	41.3	3.04	21	36.1
1967						
Jan	52.8	25.4	39.5	1.96	16	66.3
Feb	54.7	31.9	42.2	2.79	15	52.2
Mar	63.8	31.3	44.9	2.27	15	165.6
Apr	64.7	26.5	45.6	1.28	12	84.9
May	69.4	31.2	50.2	6.80	26	142.1
Jun	76.0	59.0	59.1	0.54	6	212.8
Jul	83.8	47.2	63.0	1.60	10	178.0
Aug	76.8	46.8	60.8	2.76	16	117.8
Sep	70.0	57.9	55.8	2.26	15	116.6
Oct	64.4	36.1	51.9	4.54	21	110.4
Nov	59.8	25.8	41.5	2.80	12	50.4
Dec	56.0	25.0	40.5	1.76	14	52.8
1968						
Jan	57.2	25.4	40.0	2.65	16	36.1
Feb	52.0	24.8	35.6	1.68	10	60.5
Mar	67.2	28.6	43.6	2.51	18	98.2
Apr	68.2	24.9	47.1	1.86	14	152.3
May	73.2	37.1	49.5	3.26	17	100.9
Jun	80.6	40.1	58.7	2.16	15	174.1
Jul	80.1	47.9	58.8	4.43	11	69.9
Aug	80.8	47.2	60.0	1.64	12	99.9
Sep	72.8	44.1	57.1	6.11	18	74.1
Oct	65.9	40.7	54.1	2.50	15	62.8
Nov	55.2	30.3	42.9	3.78	19	29.6
Dec	49.2	27.5	37.8	2.08	14	28.3
1969						
Jan	55.6	29.5	42.1	3.34	14	28.8
Feb	45.2	19.2	32.7	3.97	19	63.5
Mar	52.4	25.5	36.4	4.71	14	48.1
Apr	68.1	30.5	45.1	3.00	16	147.5
May	67.2	34.0	52.5	3.90	23	99.2
Jun	75.7	38.2	57.6	2.41	12	253.4
Jul	85.5	48.1	62.9	2.07	10	227.3
Aug	81.4	45.0	61.1	1.91	15	129.0

Table 1 (continued)

Month	Temperature (deg. F)			Rainfall		Sun
	Max.	Min.	Mean	Inches	Days	Hrs.
1969						
Sep	72.9	41.1	56.9	1.53	14	75.4
Oct	76.6	41.2	54.9	0.96	12	76.9
Nov	59.2	26.2	40.8	6.44	19	65.0
Dec	53.3	28.3	39.1	2.86	17	30.5
1970						
Jan	49.8	22.4	38.7	3.72	23	36.8
Feb	50.8	23.0	36.2	4.10	21	88.5
Mar	54.1	20.7	38.7	2.17	23	114.6
Apr	63.9	28.2	44.4	4.19	21	134.3
May	73.3	40.0	55.2	0.67	8	160.2
Jun	82.6	45.3	61.7	0.95	6	238.9
Jul	86.5	47.2	59.8	1.54	13	153.2
Aug	80.1	46.9	61.3	1.68	11	147.6
Sep	77.6	45.9	58.5	1.68	10	122.6
Oct	66.2	38.2	51.5	1.89	15	101.0
Nov	60.4	30.7	45.1	5.57	27	48.4
Dec	52.8	28.5	39.9	1.59	18	45.7

Month	Temperature (deg. F)			Rainfall		Sun
	Max.	Min.	Mean	mm	Days	Hrs.
1971						
Jan	54.6	23.0	40.1	69.3	20	39.9
Feb	51.6	27.7	41.1	26.7	10	63.3
Mar	54.7	29.3	41.8	47.2	19	79.8
Apr	69.0	33.2	45.8	84.1	10	70.0
May	70.5	33.1	53.0	63.5	12	207.3
Jun	72.2	43.1	54.1	66.7	16	131.5
Jul	81.8	45.5	63.4	86.3	9	207.2
Aug	73.2	45.7	60.0	102.2	16	121.1
Sep	75.1	42.1	58.0	31.2	6	148.0
Oct	74.1	34.9	53.1	92.8	9	139.3
Nov	62.3	30.1	42.1	62.6	13	78.5
Dec	57.8	32.8	44.3	51.4	11	48.2

Table 1 (continued)

Month	Temperature (deg. C)			Rainfall		Sun
	Max.	Min.	Mean	mm	Days	Hrs.
1972						
Jan	9.7	—	3.5	92.7	24	38.7
Feb	8.7	−6.2	3.9	65.9	18	28.9
Mar	18.2	−0.9	6.3	106.2	17	118.6
Apr	16.1	2.4	8.5	56.9	13	94.6
May	18.4	3.2	10.6	56.6	20	135.4
Jun	19.4	4.9	12.1	83.1	22	127.8
Jul	26.1	6.6	15.5	55.7	12	148.4
Aug	23.3	7.8	15.4	61.2	12	180.8
Sep	20.5	3.8	11.8	50.6	7	80.4
Oct	18.2	2.2	10.3	14.5	8	61.7
Nov	17.1	−0.7	7.0	84.9	18	68.6
Dec	13.0	0.1	5.2	80.3	14	22.7
1973						
Jan	10.8	−2.5	4.8	39.8	9	21.7
Feb	11.1	−5.9	4.8	65.0	13	91.7
Mar	17.2	0.2	7.1	15.2	7	127.3
Apr	16.3	−6.6	7.5	77.5	14	123.8
May	23.2	0.9	11.3	59.5	16	150.6
Jun	25.0	5.1	15.4	44.0	7	204.2
Jul	25.2	7.5	15.9	200.6	14	115.6
Aug	28.4	9.1	16.3	58.0	10	161.9
Sep	24.8	5.6	14.2	58.1	15	108.8
Oct	17.2	−0.2	9.2	63.5	14	75.7
Nov	14.8	−3.9	6.05	28.8	12	89.7
Dec	11.0	−5.2	4.8	51.6	18	41.8
1974						
Jan	12.6	−3.0	5.5	79.3	22	49.4
Feb	12.2	0.1	5.5	67.3	19	55.5
Mar	15.3	−2.0	5.4	39.8	15	95.0
Apr	16.8	0.9	7.5	12.3	4	120.5
May	20.3	2.9	11.2	39.5	11	195.1
Jun	24.7	6.5	13.7	48.5	13	184.3
Jul	23.3	8.0	15.2	67.6	16	164.0
Aug	23.2	5.9	15.5	81.0	12	180.1
Sep	22.1	3.9	12.1	90.8	24	138.7
Oct	13.4	0.7	7.6	55.9	19	55.9
Nov	12.6	−1.3	6.2	121.1	25	47.5
Dec	13.9	−1.0	7.5	71.1	25	43.2

Table 1 (continued)

Month	Temperature (deg. C)			Rainfall		Sun
	Max.	Min.	Mean	mm	Days	Hrs.
1975						
Jan	12.0	−0.4	6.4	73.2	22	32.2
Feb	11.4	−0.8	4.1	15.8	9	25.0
Mar	10.1	−1.5	4.3	58.8	18	86.4
Apr	20.5	−2.0	8.6	46.9	21	132.8
May	22.1	1.5	9.7	86.2	8	187.9
Jun	27.3	2.5	14.9	11.9	4	253.0
Jul	28.1	9.9	18.0	81.9	18	170.9
Aug	32.6	9.9	19.3	51.2	10	236.1
Sep	22.7	2.6	13.4	34.6	17	138.5
Oct	16.2	2.6	9.9	27.2	10	91.2
Nov	12.5	−1.3	6.1	28.6	16	70.6
Dec	11.9	−1.7	5.9	43.0	8	49.8
1976						
Jan	11.8	−3.9	5.4	93.8	17	41.4
Feb	14.7	−1.9	4.4	33.7	19	45.0
Mar	13.6	−2.0	4.6	29.5	13	108.8
Apr	16.9	−0.5	8.1	16.2	6	116.1
May	21.6	3.3	11.9	79.9	17	148.6
Jun	30.3	6.2	17.2	16.6	7	250.2
Jul	31.5	8.7	18.6	16.9	5	285.2
Aug	28.1	8.5	17.5	17.3	1	215.6
Sep	22.1	5.5	13.0	134.9	18	85.5
Oct	18.5	2.7	10.3	128.7	27	58.7
Nov	10.7	−1.0	6.0	32.4	15	58.7
Dec	7.1	−3.1	2.1	75.9	23	54.2
1977						
Jan	11.5	−4.2	2.5	113.8	18	45.9
Feb	10.1	−4.1	3.9	201.4	24	53.5
Mar	15.7	−1.2	6.4	59.7	21	79.0
Apr	16.0	−1.2	7.2	45.0	19	179.3
May	24.3	1.7	10.4	43.5	9	207.6
Jun	25.5	2.5	12.4	65.4	14	182.0
Jul	26.6	6.0	15.7	9.5	7	178.5
Aug	26.8	6.6	15.3	69.2	13	137.7
Sep	23.2	6.2	13.2	35.2	11	108.8
Oct	17.0	3.9	11.1	45.4	17	87.4
Nov	15.7	−1.5	5.9	112.0	19	80.9
Dec	14.7	−3.8	5.6	82.5	20	30.7
1978						
Jan	8.7	−3.7	3.0	99.4	23	52.4
Feb	11.9	−6.6	1.9	55.4	19	33.5

Table 1 (continued)

Month	Temperature (deg. C)			Rainfall		Sun
	Max.	Min.	Mean	mm	Days	Hrs.
1978						
Mar	16.8	−1.5	6.8	73.2	20	142.5
Apr	14.2	−2.9	6.2	47.3	18	89.9
May	26.2	4.8	11.8	34.7	14	180.7
Jun	26.5	6.2	13.6	90.0	14	174.4
Jul	26.0	7.9	14.6	58.8	14	150.7
Aug	25.2	6.8	15.2	48.5	20	126.9
Sep	22.4	7.4	14.1	48.8	12	126.2
Oct	22.8	3.9	12.1	9.9	7	88.4
Nov	16.4	−3.6	8.7	50.0	12	71.4
Dec	12.8	−5.9	3.0	207.9	23	24.9

The data in Table 1 are extracts from the daily records of the Weston Park Weather Station, Sheffield, for the 'High Summer'* periods of the years 1974–78.

In the record, wind direction and force were as measured at 09.00 hours GMT and wind force is on the Beaufort scale. Rainfall is measured in mm and is the total at 09.00 GMT for the previous 24 hours. Sunshine is similarly the total duration of sunshine during the 24 hours up to 09.00 GMT.

(i) Summarize wind directions and strengths separately in tabular and/or graphical form. Is there any evidence: (a) that the wind blows more often from some directions than from others; and (b) that it blows more strongly from some directions (say northerly and easterly) than from others (say southerly and westerly)?

Write a paragraph suitable for insertion in a local newspaper describing your findings.

(ii) Investigate whether there is a connection between rainfall and sunshine on the one hand and wind on the other.

* The 'High Summer' period, from 18 June to 9 September, has been identified by H. Lamb (1950) as a period over which weather in the British Isles maintains a reasonably stable homogeneous pattern.

* * *

Level: 3 *Duration:* 2/4 weeks *Type:* Individual

Comment: Students unfamiliar with circular data might welcome a reference to descriptive circular statistics: sections of Mardia (1972), for example. A possible pitfall for those who use standard hypothesis tests uncritically in part (i) is the day-to-day dependence of the observations; some thought needs to be given to whether the degree of dependence present is large enough to cast doubt on any conclusions drawn from such tests. In part (ii) various possible connections might be investigated, but, unless time is very plentiful, it would be reasonable to limit attention to only one – say the connection between rainfall alone and wind, or between sunshine alone and wind. If the latter is chosen, another explanatory variable – date – might be thought relevant to the analysis.

Table 1 Daily wind speed, wind direction, rainfall and hours of sunshine: Weston Park,
Sheffield High, Summers (18 June–9 September) 1974–78
(Reproduced by kind permission of Sheffield City Museums)

Date	Wind force & direction	Rainfall (mm)	Sun (hrs)	Date	Wind force & direction	Rainfall (mm)	Sun (hrs)
1974							
18 Jun	NW4	0.4	9.6	1 Aug	WNW2	TR	5.6
19	SSW1	—	2.9	2	W3	—	12.2
20	CALM	—	11.8	3	S2	—	6.9
21	CALM	—	14.5	4	NNE2	—	0.2
22	NE2	—	9.1	5	CALM	—	12.2
23	NNW2	—	—	6	S2	—	10.4
24	NNE2	—	1.4	7	SSE1	0.7	0.4
25	NE2	—	4.7	8	CALM	24.4	0.3
26	NNE2	0.2	0.1	9	NW3	8.5	3.4
27	NNE3	—	—	10	SSW3	0.1	3.3
28	SE1	4.9	—	11	WNW3	12.0	6.2
29	CALM	6.9	2.7	12	WSW3	—	8.0
30	CALM	7.0	0.1	13	CALM	5.5	0.4
				14	CALM	1.1	—
1 Jul	WNW4	—	8.5	15	WSW3	0.4	9.2
2	S2	10.5	2.8	16	SW2	—	12.9
3	NW4	2.8	7.2	17	W3	—	5.7
4	CALM	18.0	—	18	NNW1	0.2	6.5
5	W3	—	2.4	19	CALM	—	6.8
6	WNW5	TR	11.8	20	S2	—	11.5
7	WNW2	—	7.5	21	SW2	—	9.0
8	WSW2	—	2.8	22	W3	—	4.0
9	WNW3	1.6	12.3	23	SW2	TR	6.6
10	SSW2	2.0	0.4	24	WSW3	—	4.8
11	WSW3	0.2	3.5	25	SSW3	12.2	3.6
12	WSW3	—	2.5	26	CALM	3.1	2.0
13	CALM	0.2	2.0	27	NW3	TR	11.8
14	W3	5.5	8.5	28	S1	TR	10.8
15	SSW3	5.9	0.7	29	SE2	TR	1.8
16	WSW2	0.5	5.3	30	NNE2	3.4	0.6
17	NNE3	0.2	7.5	31	CALM	9.4	3.0
18	WNW3	—	9.5				
19	CALM	0.6	0.1	1 Sep	SSW1	1.1	4.5
20	WNW3	—	11.7	2	SE2	20.1	1.0
21	WSW3	—	13.6	3	SW4	0.1	5.2
22	SW3	0.9	6.2	4	SSW3	2.8	0.1
23	W3	—	6.3	5	SE1	6.0	1.2
24	NNW2	—	9.3	6	SW1	9.6	7.8
25	NW3	0.1	7.5	7	SSW4	9.4	1.4
26	NW1	—	1.7	8	WSW4	TR	6.9
27	WSW1	—	1.0	9	W3	—	7.2
28	WSW2	0.1	5.8				
29	WSW4	0.7	1.9				
30	CALM	14.6	—				
31	SSW2	3.2	3.7				

Table 1 (continued)

Date	Wind force & direction	Rainfall (mm)	Sun (hrs)	Date	Wind force & direction	Rainfall (mm)	Sun (hrs)
1975							
18 Jun	SW3	—	5.7	1 Aug	ENE2	—	8.6
19	SSW3	—	4.5	2	SSW1	—	12.8
20	W2	—	2.3	3	SSW1	—	13.0
21	NNE1	—	6.5	4	ENE1	—	13.7
22	NNE2	—	6.4	5	NE2	1.3	3.0
23	NNE2	—	15.5	6	SSW3	—	10.0
24	NNW2	TR	2.9	7	ENE1	—	11.4
25	SSE1	—	15.4	8	NE1	17.4	9.2
26	WNW2	—	9.1	9	NE1	—	4.2
27	NNE3	—	0.1	10	CALM	—	8.7
28	NNE2	—	7.0	11	NE1	—	8.8
29	NE2	—	11.6	12	NNE1	5.2	6.2
30	NNW3	—	12.3	13	NNE1	TR	8.2
				14	ESE1	5.8	4.7
1 Jul*	NNW6	—	6.8	15	WNW2	1.5	3.6
2	NE1	—	13.5	16	WNW3	—	10.0
3	SE3	—	12.4	17	WNW2	—	5.4
4	NE5	—	3.3	18	ENE1	—	3.7
5	CALM	—	8.0	19	S2	0.5	4.1
6	CALM	—	14.0	20	WSW4	—	10.4
7	E4	0.9	4.0	21	NW1	TR	5.0
8	NE7	TR	0.4	22	NW3	—	6.3
9	CALM	5.0	2.0	23	NW2	0.4	5.1
10	S11	5.6	2.5	24	WNW3	3.3	7.3
11	SSW10	0.5	3.0	25	NNE1	—	11.7
12	SW1	2.4	1.0	26	WNW1	—	8.4
13	SW5	19.2	2.3	27	CALM	—	10.8
14	S1	6.9	6.1	28	W1	—	12.7
15	WSW9	3.7	6.2	29	CALM	1.9	9.1
16	WSW12	0.3	5.7	30	CALM	9.4	—
17	SSE2	21.4	2.0	31	NW2	TR	—
18	CALM	5.7	1.8				
19	SSW5	1.4	1.7	1 Sep	NNW1	TR	12.0
20	CALM	0.4	3.5	2	CALM	TR	9.6
21	WSW7	—	7.3	3	NW2	—	3.6
22	SW15	2.3	—	4	NNW1	0.5	8.2
23	W18	3.3	10.0	5	WSW1	0.1	1.9
24	WNW15	1.9	7.1	6	SW1	TR	1.9
25	NW8	0.6	5.8	7	CALM	—	0.3
26	W2	0.4	1.5	8	S3	6.5	3.9
27	CALM	—	13.5	9	CALM	0.4	5.5
28	W4	—	13.9				
29	NNE1	—	6.4				
30	W7	—	5.7				
31	N6	—	—				

(* Wind force for July given inadvertently in mph, not Beaufort number)

Table 1 (continued)

Date	Wind force & direction	Rainfall (mm)	Sun (hrs)	Date	Wind force & direction	Rainfall (mm)	Sun (hrs)
1976							
18 Jun	SW1	TR	0.4	1 Aug	W3	—	2.2
19	W2	0.2	0.4	2	W1	—	—
20	NW3	0.2	9.7	3	WNW2	—	6.2
21	S1	—	6.9	4	W2	—	7.3
22	WSW1	—	12.5	5	CALM	—	8.0
23	CALM	—	8.1	6	CALM	—	2.1
24	S2	—	15.0	7	CALM	—	0.1
25	CALM	—	14.4	8	CALM	—	3.1
26	CALM	—	15.0	9	CALM	—	11.1
27	CALM	—	16.0	10	CALM	—	11.4
28	CALM	—	15.6	11	CALM	—	10.5
29	CALM	—	15.2	12	CALM	—	9.7
30	CALM	—	15.1	13	CALM	—	7.2
				14	CALM	—	0.4
1 Jul	ESE1	—	13.4	15	CALM	—	12.2
2	CALM	—	15.3	16	CALM	—	9.0
3	CALM	0.6	12.9	17	CALM	—	11.8
4	CALM	0.7	9.9	18	NE1	—	7.0
5	N1	—	12.8	19	CALM	—	10.6
6	N1	—	12.1	20	N1	TR	12.8
7	NNE1	—	14.2	21	CALM	—	7.0
8	ESE1	—	14.5	22	E2	—	13.0
9	SSW1	—	3.8	23	CALM	—	11.1
10	CALM	—	11.4	24	CALM	—	12.6
11	CALM	TR	13.6	25	CALM	—	12.3
12	SSE1	11.7	1.2	26	CALM	—	12.1
13	SSW1	—	4.8	27	N2	—	0.2
14	WSW3	TR	9.8	28	NE2	17.3	0.5
15	S3	1.9	1.3	29	CALM	TR	3.2
16	CALM	—	8.6	30	N1	TR	0.1
17	CALM	—	11.8	31	N1	TR	0.8
18	SW1	TR	12.1				
19	NW4	0.1	4.4	1 Sep	W1	4.2	—
20	NW1	TR	8.5	2	NNW2	—	6.4
21	N2	—	7.8	3	WNW2	—	8.3
22	WNW2	—	13.5	4	NE2	—	2.1
23	WNW2	—	6.1	5	CALM	—	7.9
24	NE (DEF)	—	6.4	6	WNW1	—	2.6
25	NE1	—	12.7	7	CALM	—	9.2
26	CALM	—	1.1	8	CALM	2.6	0.2
27	CALM	—	12.1	9	NW3	0.8	2.7
28	N1	—	5.7				
29	W3	—	12.0				
30	NW2	—	2.7				
31	NNW1	1.9	8.7				

Table 1 (continued)

Date	Wind force & direction	Rainfall (mm)	Sun (hrs)	Date	Wind force & direction	Rainfall (mm)	Sun (hrs)
1977							
18 Jun	NNE2	—	—	1 Aug	CALM	—	14.5
19	NNE2	—	6.3	2	CALM	—	6.2
20	NW1	—	3.7	3	NW1	—	7.3
21	N2	—	12.1	4	SSW3	0.8	6.8
22	NNW1	—	15.6	5	WNW4	—	9.4
23	CALM	—	9.4	6	W1	TR	2.8
24	CALM	—	5.7	7	SE2	0.1	0.5
25	W1	—	0.6	8	DEF2	—	0.3
26	NW2	—	10.6	9	CALM	—	12.7
27	ESE3	4.1	3.3	10	CALM	—	7.7
28	S2	1.1	2.5	11	CALM	—	12.5
29	W3	TR	13.7	12	CALM	TR	1.8
30	S2	2.5	2.5	13	S1	TR	0.3
				14	SW1	5.4	—
1 Jul	SW3	0.9	0.7	15	CALM	0.2	1.3
2	SW2	—	11.4	16	NE3	TR	4.8
3	CALM	—	11.7	17	NNE3	0.5	—
4	CALM	—	13.2	18	N3	21.2	—
5	NE2	—	15.2	19	NW3	4.8	—
6	NNE2	—	12.4	20	CALM	4.1	1.7
7	N1	—	13.3	21	S1	4.1	—
8	SSE2	—	0.6	22	NNE3	—	8.3
9	NNE3	—	—	23	CALM	TR	9.3
10	NNE2	—	0.9	24	S2	11.9	—
11	N2	—	0.2	25	SE3	12.2	2.0
12	NNE1	0.1	—	26	SSW2	1.3	6.3
13	SE1	—	—	27	NNW2	0.2	0.6
14	N3	—	7.6	28	CALM	—	11.0
15	CALM	—	11.6	29	N3	2.4	5.8
16	WNW3	—	12.4	30	W1	TR	0.4
17	SSW2	3.1	0.3	31	SE1	TR	3.4
18	WNW3	—	2.6				
19	W2	0.5	6.5	1 Sep	SE3	2.6	3.6
20	WNW2	TR	4.3	2	S2	—	0.9
21	WNW2	0.2	4.8	3	WSW3	—	10.4
22	CALM	—	1.7	4	WNW3	1.6	0.9
23	WSW4	TR	4.2	5	SSW3	—	7.8
24	WNW4	0.6	5.3	6	SSW4	TR	5.3
25	CALM	3.0	3.3	7	CALM	—	5.0
26	NNW2	—	8.1	8	SW1	2.5	3.6
27	NW2	—	4.9	9	NW2	0.4	6.6
28	CALM	1.1	2.2				
29	N1	—	1.4				
30	CALM	—	8.7				
31	CALM	—	8.8				

Table 1 (continued)

Date	Wind force & direction	Rainfall (mm)	Sun (hrs)	Date	Wind force & direction	Rainfall (mm)	Sun (hrs)
1978							
18 Jun	NNW1	—	11.2	1 Aug	NW1	0.4	—
19	CALM	—	12.0	2	NNW1	4.6	0.2
20	CALM	—	10.4	3	WSW3	2.2	0.7
21	SSE3	6.4	0.4	4	SSW3	2.1	5.0
22	SSW1	0.3	8.3	5	SSW1	1.5	2.7
23	CALM	0.4	0.6	6	SSW2	2.5	1.1
24	SW2	0.1	2.4	7	WNW2	3.5	0.9
25	SW2	—	5.2	8	W1	13.6	0.4
26	SSW3	1.2	2.3	9	WNW2	0.2	2.7
27	WSW3	5.3	3.3	10	CALM	—	0.1
28	SW2	1.8	0.1	11	SE2	—	9.8
29	WSW1	9.1	—	12	SE2	0.6	2.1
30	W2	4.7	6.1	13	WSW2	1.4	9.3
				14	SE3	2.8	1.1
1 Jul	SSW2	6.4	—	15	SE3	0.2	8.0
2	WSW3	2.3	0.1	16	S3	0.3	4.7
3	SW3	0.3	3.7	17	SW4	—	8.7
4	WSW2	3.9	3.6	18	ESE1	—	8.7
5	WNW3	1.6	—	19	SE2	0.5	13.0
6	CALM	—	1.3	20	CALM	0.6	1.4
7	W3	0.8	4.0	21	SSE3	TR	5.0
8	SSW3	0.1	3.6	22	WSW3	3.3	3.6
9	SW3	TR	12.3	23	W1	—	1.7
10	CALM	—	2.8	24	WSW1	—	1.9
11	CALM	TR	6.2	25	W1	—	10.5
12	CALM	—	3.8	26	CALM	—	5.7
13	CALM	—	6.6	27	CALM	—	3.9
14	NW2	—	9.3	28	SW1	—	0.3
15	CALM	—	7.5	29	SW2	2.2	3.7
16	CALM	—	7.9	30	NNW2	2.7	9.5
17	CALM	—	11.0	31	S1	3.3	0.5
18	WSW2	TR	5.5				
19	W2	TR	0.9	1 Sep	WNW1	TR	2.3
20	W3	TR	1.5	2	W1	—	7.6
21	WSW2	TR	4.5	3	WSW2	TR	4.2
22	SSE3	3.4	4.3	4	CALM	TR	—
23	SSE3	1.3	0.8	5	CALM	4.1	—
24	S2	—	6.5	6	CALM	4.5	0.1
25	SSE2	7.5	10.8	7	S1	TR	2.2
26	SE1	1.4	1.7	8	ESE2	2.1	7.4
27	SE1	6.6	7.2	9	ESE2	1.5	0.8
28	S3	—	11.0				
29	CALM	11.2	12.3				
30	WNW1	3.0	—				
31	NW2	9.0	—				

Use the data in Project 21, 'Wind, rain, and sun' to investigate whether there are any simple patterns of day-to-day dependence in the weather that might enable you to give some useful forecasting rules – such as 'if it rained yesterday there is a chance (?) that you will need your umbrella today, and if it didn't there's a chance (!) that you will'.

*   *   *

Level: 3   *Duration:* 2 weeks   *Type:* Individual

Comment: This could be a sequel to Project 21. As with other projects concerning weather, if local data can be found, so much the better. Forecasting for rain is emphasized above, but the same question makes sense in relation to sunshine too (but then, in Sheffield at least, has less immediate personal relevance.)

The references below are concerned with different aspects of the problem of monthly weather forecasting.

Write a critical review of each paper, explaining:

(i) the purpose of the paper;
(ii) its assumptions, method(s) and conclusions; and including, if appropriate,
(iii) a discussion of any limitations; and
(iv) any further analysis that seems desirable to you.

Comment, in the light of your reviews, on the state of monthly weather forecasting as reflected by these papers.

GORDON, A. (1974) Accuracy of weather forecasts, *Nature*, **252**, 294–295.
GREEN, F. H. W. (1975) The February–June weather relationship in north-west Europe, *Nature*, **253**, 522–523.

*   *   *

Level: 2/3   *Duration:* 2/3 weeks   *Type:* Individual

Comment: For a discussion of this particular project see Section 4.6.3.

In a routine eyesight examination of 8-year-old Glasgow school children in 1955, the children were divided into two categories; those who wore spectacles (A), and those who did not (B). As a result of the test, visual acuity was classed as good, fair, or bad. The children wearing spectacles were tested with and without them. The results are given in Table 1.

Table 1 Visual acuity of Glasgow school children (1955)

Category	A, with specs.		A, without specs.		B	
	Boys	Girls	Boys	Girls	Boys	Girls
Good	157	175	90	81	5908	5630
Fair	322	289	232	222	1873	2010
Bad	62	50	219	211	576	612
Total	541	514	541	514	8357	8252

What conclusion can be drawn from these data, regarding (a) sex difference in eyesight, (b) the value of wearing spectacles?

The figures for 8-year-old boys and girls for the years 1953 and 1954 were not kept separately for the sexes. They are shown in Table 2.

Table 2 Visual acuity of Glasgow school children (1953, 1954)

Category	A, with specs.		A, without specs.		B	
	1953	1954	1953	1954	1953	1954
Good	283	328	152	173	8743	10,511
Fair	454	555	378	443	3212	3565
Bad	84	78	290	345	1015	1141
Total	820	961	820	961	12,970	15,217

Are there, in your opinion, any signs of changes in the 8-year-old populations of visual acuity with time? Do you think it possible that your conclusions might be vitiated by the pooling of the frequencies for both sexes in 1953 and 1954?

(London BSc 1958)

*   *   *

Level: 3/2   *Duration:* 1/2 weeks   *Type:* Individual

Comment: The challenge of this project lies not in assembly of relevant data and recognition of the substantive problem – the questions of interest are already posed, the data are given and there is little hope of obtaining more – but rather in the finding of valid statistical analyses to help answer the questions, and the judging of what conclusions are reasonable even when the information on which they are based is incomplete. An initial consideration is whether the question refers to a population or a sample. A thorough discussion would look at both possibilities, but the teacher might wish to avoid confusion by suggesting that students concentrate on the latter. (There is still then room for discussion about what kind of sample 8-year-old Glasgow school children constituted in 1955, and from what population.) If significance tests are thought appropriate it is perhaps fair to warn students to exercise care in applying them: tables of frequencies are not necessarily contingency tables.

It is obvious that lengths of features (such as roads, rivers and footpaths) in reality must be somewhat greater than the lengths derived from maps, since not all irregularities, such as bends, can be reproduced in a map. Investigate the relationship between lengths derived from maps and lengths 'on the ground'.

*   *   *

Level: 1/2/3   *Duration:* 2–4 weeks   *Type:* Group

Comment: Different types of feature, with different amounts of irregularity, probably have different relationships; certainly different scales of map will represent irregularities with varying degrees of success and a possible variant of the project would confine itself to data collection from maps alone.

Errors of length resulting from the use of maps are in fact of various kinds: those caused by the projection, those arising from measurement error, and those which occur during the making, distributing, and storing of a map – including shrinkage of the paper, the use of symbols which, for the sake of visibility, are too large, and which cannot, therefore, represent all the fine details of, for example, the course of a real footpath or river. (It is quite arguable, in any case, that for all real geographical features there is no such thing as a true length.) Certainly only some of these aspects can be considered in a project of this kind, and the decision on which to treat is left to the teacher or student.

Estimate the probability that your surname will die out ultimately in your country, assuming a simple Galton–Watson process model for the way it is handed on*.

According to this model the probability of extinction is π^N, where N is the current number of males with your surname, and π is the smallest positive root of the equation $z = A(z)$, where $A(z)$ is the p.g.f. of the number of male offspring in families with your surname. You will therefore need to estimate both N and the offspring distribution.

For N, one method might be to sample from telephone directories and make adjustments for the proportion of households with telephones and the number of males per household.

For the offspring distribution, data may be found in an appropriate government statistical publication.

Discuss

- the assumptions that you have had to make in the above;
- limitations of the model;

and try

- to suggest ways of obtaining a better estimate.

* See, for example, Feller (1968), Sections XII.4, XII.5.

*   *   *

Level: 3   *Duration:* 4 weeks   *Type:* Individual

Comment: Of the two parts of this project the first (estimation of N) is the less . . . artificial, but useful lessons can be learnt from the other part too. In the UK many large libraries hold a full set of national telephone directories. Regional variations in surname frequencies demand that thought be given to the way sampling is done. Information on numbers of households with telephones and on demographic questions may be found in *Social Trends* and in the *Registrar-General's Reports* respectively. Even with extensive data there will still be a need to make assumptions. Part of the point of the project is to give an opportunity for the student to exercise judgement about what is plausible, and to quantify the influence of different assumptions. The focusing of attention on the Galton–Watson model should engage the critical faculties and perhaps stimulate thought about other models.

In the 1981 French Presidential Election there were two stages: Round 1 on 26 April, and Round 2 on 10 May. Ten candidates stood in Round 1:

François Mitterand	A
Valéry Giscard-d'Estaing	B
Jacques Chirac	C
Georges Marchais	D
Brice Lalonde	E
Michel Crépeau	F
Arlette Laguiller	G
Michel Debré	H
Marie-France Garaud	J
Huguette Bouchardeau	K

In the second round there were only two candidates, namely A and B, who had achieved first place and second place in Round 1. Each received considerably more votes in Round 2 than in Round 1, the increases presumably coming from electors who had voted for other candidates or had abstained in the first round. It is of interest to know how A's and B's increased votes were made up from the possible sources.

Table 1 shows the results of the election for 24 Departments of Metropolitan France, chosen systematically (every fourth Department) from an alphabetical list. All table entries represent thousands. The first column, headed 'Electeurs inscrits', gives the number of registered electors. The exact numbers of registered electors fell slightly between the two rounds but never by more than about 0.2%.

Examine the data in any way that seems appropriate to you (formal or informal or both) and summarize their salient features. Report in particular on what light you think the data throw on the question at the end of the first paragraph above, that is, where did the new votes which A and B gained between Rounds 1 and 2 come from?

*   *   *

Level: 2/3   *Duration:* 1/2 weeks   *Type:* Individual/small group

Comment: This is suitable at several levels, because several different approaches at different levels of sophistication can each lead to useful conclusions. See Section 4.6.4.

Table 1 The 1981 French Presidential Election: votes in 24 Departments

	Electeurs inscrits	Round 1										Round 2	
		A	B	C	D	E	F	G	H	J	K	A	B
1. Ain	260	51	64	36	23	9	5	4	4	3	3	105	114
2. Alpes (Hautes)	75	14	17	9	9	3	1	2	1	1	1	32	31
3. Ariège	107	27	18	13	17	2	2	2	1	1	1	57	33
4. Bouches-du-Rhône	1036	191	204	119	205	29	13	13	10	10	6	466	364
5. Charente-Maritime	367	71	76	47	37	8	34	5	4	4	2	163	142
6. Côtes-du-Nord	396	93	90	57	54	13	5	9	4	3	5	193	155
7. Drôme	257	57	55	31	30	10	4	5	4	3	3	116	99
8. Finistère	595	132	149	95	49	21	9	11	6	5	10	249	259
9. Gironde	735	195	137	98	83	20	16	13	13	8	5	356	261
10. Indre	181	34	39	28	28	4	3	4	3	2	1	82	72
11. Landes	219	62	47	31	26	5	3	3	3	2	1	107	84
12. Loire-Atlantique	653	149	156	94	49	23	15	13	10	7	8	274	275
13. Lozère	58	10	18	9	4	2	0	1	1	1	1	20	29
14. Marne (Haute)	145	32	33	20	15	4	2	3	2	2	1	63	59
15. Morbihan	414	86	117	65	33	14	6	8	5	4	4	162	190
16. Oise	416	87	88	59	62	13	7	10	6	5	3	192	160
17. Pyrénées-Atlantique	391	90	91	66	33	12	6	6	5	4	3	165	168
18. Rhin (Haut)	413	75	125	58	19	17	6	8	6	6	4	138	204
19. Sarthe	346	72	87	49	40	10	6	8	4	3	3	149	145
20. Seine-Maritime	783	171	181	91	123	24	13	18	10	7	6	370	297
21. Sèvres (Deux)	240	54	66	34	16	8	7	5	3	4	2	98	108
22. Val D'Oise	533	111	100	74	81	22	12	10	7	7	6	252	192
23. Vendée	336	61	105	59	19	10	11	6	5	4	3	115	176
24. Yonne	216	44	52	31	24	7	4	4	3	3	2	91	91

Votes (thousands)

Source: Le Monde 28 April and 12 May 1981. Reproduced by kind permission of the Editor, *Le Monde*.

Digestive System of Sheep

As part of an investigation into the digestive system of sheep, 4000 indigestible plastic beads were placed into the rumen of a sheep. The sheep was fed every 6 hours and its faeces were collected and analysed for bead passage.

The counts of the number of beads excreted in each time interval are given in Table 1 together with the number of beads still retained in the gastrointestinal tract (i.e. not yet excreted) starting four days after the introduction of the beads. Towards the end of the experiment some of the counts were made over two or three six-hour periods.

Time (days)	No. of beads retained	No. of beads excreted	Time (days)	No. of beads retained	No. of beads excreted
			13.00	3,202	26
4.25	3,989		0.25	3,179	23
.50	3,970	19	.50		
.75	3,954	16	.75	3,127	52
5.00	3,935	19	14.00	3,105	22
.25	3,931	4	.25	3,062	43
.50	3,905	26	.50	3,011	51
.75	3,888	17	.75	2,976	35
6.00	3,872	16	15.00		35
.25	3,864	8	.25	2,941	
.50	3,832	32	.50	2,930	11
.75	3,826	6	.75		
7.00	3,813	13	16.00		19
.25	3,795	18	.25	2,911	
.50	3,778	17	.50	2,891	20
.75	3,761	17	.75	2,866	25
8.00	3,705	56	17.00	2,839	27
.25	3,662	43	.25	2,821	18
.50	3,629	33	.50		
.75	3,622	7	.75	2,796	25
9.00	3,599	23	18.00	2,774	22
.25	3,589	10	.25	2,766	8
.50	3,570	19	.50	2,758	8
.75	3,563	7	.75	2,744	14

Time (days)	No. of beads retained	No. of beads excreted	Time (days)	No. of beads retained	No. of beads excreted
10.00	3,556	7	19.00	2,736	8
.25	3,542	14	.25	2,727	9
.50	3,531	11	.50	2,714	13
.75	3,503	28	.75		
11.00	3,473	30	20.00	2,701	13
.25	3,450	23	.25	2,696	5
.50	3,391	59			
.75	3,374	17			
12.00	3,347	27			
.25	3,318	29			
.50	3,250	68			
.75	3,228	22			

Analyse the data with a view to finding and evaluating a suitable statistical model to describe the process.

(London BSc 1974)

*   *   *

Level: 3   *Duration:* 2 weeks   *Type:* Individual/small group

Comment: It is essential to do some reading to be clear about the general structure of a sheep's digestive system before beginning this. A sophisticated analysis might be based on a Markov compartmental model, but simple re-expression and/or smoothing might provide a useful, if more naive, alternative.

PROJECT 29   *Future Numbers in Schools and Universities*

Derive projections of the country's school and university populations for 1990, 1995, and 2000.

(You will need to take account of trends in the birthrate, etc., for which [in the UK] *Annual Abstracts* or *The Registrar General's Reports* are possible sources. Participation rates for various types of education may be found in *Education Statistics* and elsewhere.)

*   *   *

Level: 3   *Duration:* 4 weeks   *Type:* Individual/small group

Comment: This project is similar in type to Project 30, 'Prison population' and Project 31, 'The demand for mathematics teachers'. Although technical demands are not great, some maturity is called for. Some guidance on sources of official statistics might be desirable.

PROJECT 30   *Prison Population*

Suppose that you are a relatively junior civil servant, whose superior is required to consider the prison-building programme for the next few years. You are instructed to prepare a report which attempts to predict the prison population in 15 years' time; the determination of policy is not your responsibility, though you should, of course, be awake to problems of this kind. Your report should describe the basis of your predictions and any doubts about their validity, in a way appropriate for a numerate but not statistically trained reader.

$$* \quad * \quad *$$

Level: 3   *Duration:* 2/4 weeks   *Type:* Individual

The Demand for Mathematics Teachers

Suppose that you are a relatively junior civil servant whose department is required to consider what provision might be needed for the training of teachers in the medium-term future. You are instructed to prepare a report which attempts to predict the annual demand for secondary-school mathematics teachers in 15 years' time. The determination of policy is not your responsibility but you should of course be alive to the effect of policy, and changes in it, on problems of this kind.

Your report should describe the basis of your predictions, and any doubts about their validity, in a way intelligible to a numerate but not statistically trained reader.

*   *   *

Level: 3   *Duration:* 3 weeks   *Type:* Individual/small group

Comment: There is, of course, no right answer – though there are plenty of wrong ones. A sensible approach is all that can be demanded. To get numerical answers requires some fairly heroic assumptions about demand and about the turnover of mathematics teachers.

The references below deal, in varying degrees and in varying manner, with the problem of describing the outcomes of ball games (particularly association football). Write a short ciritical review of each reference, concentrating particularly on the assumptions used by the authors in modelling ball games. In the light of these articles, describe the approach you would take towards the problem of setting up a model for analysing and forecasting the results of association football matches.

MORONEY, M. J. (1951) *Facts from Figures*. Pelican, pp. 96–107.

REEP, C., and BENJAMIN, B. (1968) Skill and chance in association football. *J. Roy. Statist. Soc.* A, **131**, 581–585.

REEP, C., POLLARD, R., and BENJAMIN, B. (1971) Skill and chance in ball games. *J. Roy. Statist. Soc.* A, **134**, 623–629.

HILL, I. D. (1974) Association football and statistical inference. *Applied Statistics*, **23**, 203–208.

THOMPSON, M. (1975) On any given Sunday: fair competitor orderings with maximum likelihood methods. *J. Amer. Statist. Assoc.*, **70**, 536–541.

$$*\quad*\quad*$$

Level: 3   *Duration:* 2 weeks   *Type:* Individual

(This project is due to Dr M. J. Maher, to whom we are most grateful.)

Reference (1) below describes a study of the diet of a small American spider. Interpretation of the experimental results in the article relies quite heavily on statistical arguments, some of which have been criticized in reference (2). The author of the original article replied to the criticism in reference (3).

Write critical reviews of the statistical content of the three references, explaining where appropriate:

(i) the purposes of the statistical part of the article or letter;
(ii) its assumptions, methods, and conclusions;
(iii) a discussion of any limitations; and including if necessary
(iv) any further analysis that seems desirable to you.

In the light of your reviews, which of the parties in the controversy, if either, would you say appears to have the stronger case?

(1) GREENSTONE, M. H. (1979) Spider feeding behaviour optimizes dietary essential amino acid composition. *Nature*, **282**, 501–503.
(2) GREENWOOD, J. J. D. (1981) Does spider feeding behaviour optimize dietary essential amino acid composition? Letter to Editor, *Nature*, **290**, 165.
(3) GREENSTONE, M. H. (1981) Rejoinder. *Nature*, **290**, 165–166.

*   *   *

Level: 3   *Duration:* 2 weeks   *Type:* Individual/small group

Planning is beginning for a Himalayan mountaineering expedition two years hence to climb a new hard route up a peak of over 8000 m. There are to be eight climbers plus as many Sherpas as required (30 + ?). The route that the expedition will follow on the face of the mountain is pretty well established following several earlier unsuccessful expeditions, at least up to 7600 m: in particular the sites of Base Camp and of four higher camps are known in advance. The summit assault is to be mounted from camp 4 at 7590 m without the use of oxygen. It is desirable that as many of the climbers as are fit and at least one Sherpa should have the chance to make the summit bid, so the aim is to establish accommodation for four at Camp 4, in the shape of two 2-man 'assault boxes'. The logistic problem is how to organize the establishment of this and the lower camps, and the flow of supplies through them, so as to allow summit bids to be mounted as soon as possible after climbing on the mountain commences. Speed is important (a) to avoid the onset of severe winter weather (the expedition is to take place in the short window between the end of the monsoon and the start of winter), and (b) because prolonged stay at high altitude impairs the physical condition of climbers and Sherpas.

One possible solution to the logistic problem – but a poor one – would be to establish and stock each camp completely before moving forward to the next. Clearly this would be most inefficient, since, for example, it could be improved by allowing lead climbers to begin making the route to the next camp while others are building up supplies below. The question, however, is, exactly how this should be done. The expedition leader has a rough plan based on his experience of many other expeditions, but he is willing to admit that it may not be optimal, and anyway he is keen to discuss it and any other promising possibilities further, because he realizes that he will be in a better position to make the inevitable modifications on the mountain if he has explored the problem more fully beforehand. He asks you as an independent consultant to help in this. (A large bank has adopted the expedition so there is no need to worry about your fee.) As a first step, being a competitive sort of character, he suggests that you formulate as efficient a plan as you can without seeing his proposal. The two of you can then compare ideas, argue about differences, and generally discuss the problem more fully.

Planning data are given below [see Exhibit 12]. An initial guess is that 30 Sherpas/high altitude porters will be needed, and it might be reasonable to begin work on this basis. If the number is too large or too small then you should feel free to vary it: similarly with equipment. (Money for wages and equipment is not a problem.) Your plan should specify for each day of the expedition how many climbers and Sherpas are in position at each camp and how many loads are moved between camps. Bear in mind that weather may interfere with movement, so that reserves of consumables such as food and fuel are desirable at each camp. Add notes to your plan pointing out any critical points or bottlenecks.

$$* \quad * \quad *$$

Level: 3   *Duration:* 4 weeks   *Type:* Group/Individual

Comment: This is really a rather elaborate exercise in stock control, though it is doubtful whether standard stock-control techniques will be of much help. The main challenges are firstly to recognize what is required and to formulate the problem, and then to take account in detail of the large number of constraints that are imposed on movements. It might be helpful, but it is not absolutely essential, to write a computer program to speed any calculations needed. Further background information will be found in the unusually detailed appendices of Chris Bonington's account of the 1975 expedition to the SW face of Everest (Bonington, 1976), the book which suggested this project. Bonington's own approach to the planning of such an expedition was to start with the summit bid and then to work backwards to find the build-up of men and equipment needed to mount it. He allowed for bad weather by working out the best possible rate of progress and ensuring that there were enough reserves at each camp to allow climbers to sit out periods when movement was impossible. In Appendix 2 of Bonington (1976), Stephen Taylor describes a computer program used to study strategies for the 1975 expedition.

EXHIBIT 12

PLANNING DATA	
CAMPS	
Camp 4	7590 m
Camp 3	7010 m
Camp 2	6490 m
Camp 1	5910 m
Base camp	5425 m

Initially all men and equipment are at base camp.

TENTAGE

Accommodation on the high face (Camps 2–4) is in 'assault boxes' (Camp 4) and 'face boxes' (Camps 2 and 3). Lower camps may use 'super boxes' and/or conventional 3-man tents.

Accommodation	Capacity (Men)	Weight (lb)
Assault box	2	30
Face box	2	100
Super box	8	120
3-Man tent	3	30

The larger items can be broken down into smaller loads.

TACTICS

Climbers working in pairs establish the route between camps and fix ropes, which Sherpas and other climbers can then use. It is much easier to climb fixed-roped sections than to break new trail. Lead climbers establishing the route cannot at the same time carry loads.

ESTIMATED TIMINGS

(a) For making routes: estimated times are as follows.

Base to Camp 1 (through Ice Fall)	6 days
Camp 1 to Camp 2	1 day
Camp 2 to Camp 3	4 days
Camp 3 to Camp 4	9 days
For fixing rope above Camp 4 to facilitate summit bid	1 day

Timings above Camp 3 are tentative (especially that from Camp 3 to Camp 4), depending on the technical difficulty of the face.

(b) For carrying loads: a load carried from a camp to the next higher camp is counted as one day's work. It is possible to make a carry and drop back down to the starting camp on the same day. (In the same way lead climbers may start from a camp one below that above which they are making the route that day: e.g. they could set out from Camp 2, climb rapidly without loads up fixed ropes to Camp 3, and spend the rest of the day extending the route towards Camp 4.)

No allowance has been made for bad weather in these estimates.

WEATHER

Weather cannot be predicted, of course, but experience suggests that in the early post-monsoon period there may be occasional storms severe enough to preclude movement on the mountain for a day, or possibly two consecutive days; as winter approaches such storms become more prolonged and more frequent. The weather window between the monsoon and winter is usually reckoned to extend from late August until possibly mid-October.

LOADS

(a) Sherpas

        Base to Camp 2   35–38 lb

        Camp 2 to Camp 4   30 lb

(b) Climbers

        Base to Camp 2   24 lb (2/3 load)

        Above Camp 2   20 lb

All climbers carry their personal gear (sleeping bag, clothes, etc.) as well.

REST DAYS

1. Rests from carrying loads

(a) Sherpas

        Base to Camp 2   1 day in 4

        Camp 2 to Camp 4   1 day in 2

(b) Climbers

        Base to Camp 4   1 day in 2

2. Rests from extending the route (climbers only)

        Base to Camp 2   1 day in 4

CARRYING EFFICIENCY

Carries may not take place as planned because of sickness, low morale, misunderstandings, etc. Estimated percentage of loads not being taken to their destination:

    Between Base and Camp 2   10%

    Between Camp 2 and Camp 4   30%

MAKE-UP OF LOADS

(a) Non-consumables. Apart from tents, these include camp-kits (stoves, lights, etc.) radio, medical kits, foam mats, etc. Once in position at a camp they stay there. One camp-kit is needed for each tent or box and one radio and medical kit for each camp.

Weights:

        Camp-kit   20 lb

        Radio & batteries   4 lb

        Medical kit   6 lb

        Foam mats are included with tents.

(b) Consumables:

    (i) Food and fuel. This will be packed in man-day units of two types:
        High-altitude, for use at Camp 2 and above, each man-day pack weighing 4 lb.
        Standard, for use at Camp 1, weighing 5 lb.

    (ii) Climbing gear. This includes ropes, karabiners, pegs, deadmen, ice-fall ladders, marker poles, etc. Weights of estimated requirements for different sections are:

Base to Camp 1 (through Ice Fall)	300 lb
Camp 1 to Camp 2	150 lb
Camp 2 to Camp 3	100 lb
Camp 3 to Camp 4	120 lb
Above Camp 4	20 lb

When the counting of the votes in an election in a UK parliamentary constituency is finished, a candidate or his agent may ask for a recount; multiple recounts are also possible. The usual ground is that the winning candidate's margin is small, though the request might also reflect an attempt to save the potential loss of a deposit. The returning officer is not, however, obliged to order a recount: he can refuse if he thinks the request 'unreasonable'. What is 'reasonable' or 'unreasonable' in these circumstances is not laid down, but is left to the judgement of the returning officer.

Suppose a new code of practice about the conduct of elections is being prepared for circulation to returning officers, and that you as a statistician have been asked for advice on what guidelines might be offered on the question of recounts. The authors of the code are considering whether it would be possible to offer some simple rule of thumb (e.g. majority less than 2%, say) to help decide what is a 'reasonable' request for a recount.

Discuss the problem (so far as you can in the time available) with the aim of giving some definite advice on what could be included in the code of practice, or at least, if that does not seem practicable, of formulating a programme of further investigation which would lead to such advice.

Notes: 1. In such elections a single representative is elected, based on a simple majority of the (valid) votes cast.
   2. Details of the conduct of the counting at parliamentary elections are briefly:

      (i) Ballot boxes are brought to the central place of counting once polling finishes.
      (ii) Each ballot box is opened and the voting papers it contains are counted.
      (iii) Papers from all ballot boxes are mixed and then sorted into those for the various candidates, and those deemed to have been spoiled.
      (iv) Votes are counted by hand into bundles of twenty and these are then combined into one hundreds; the bundles of twenty retain their identity.

   3. Each candidate must deposit £500 prior to the election, which is returned to him provided he receives at least 5% of the total votes cast.

References

CREWE, I., and FOX, A. (1984) *British Parliamentary Constituencies and Statistical Compendium.* Faber & Faber, London.
SCHOFIELD, A. N. (1959) *Parliamentary Elections* (3rd ed.), Shaw, London; describes the law and practice of these elections.

*   *   *

Level: 3   *Duration:* 1/2 weeks   *Type:* Individual/group

Comment: This is hard, but an answer might be based on modelling the counting process – errors can occur at several steps with varying probabilities. Estimates of these probabilities are not, as far as we know, available, and some research would therefore need to be proposed to determine them if they are needed; orders of magnitude might be obtained by talking to banks, which face somewhat similar problems in counting bank-notes.

Table 1 shows some of the history of the exploration of a new petroleum province* in the early 1970s. It records, for each successful exploration well, the recoverable reserves of oil (in barrels – BBLS) attributable to the field discovered. The numbering of wells corresponds to the order in which they were drilled. Wells not listed in the table were unsuccessful.

Table 1 Field sizes of 58 discoveries out of 220 exploration drillings

Exploration well no.	Field size (oil 10^6 BBLS)	Exploration well no.	Field size (oil 10^6 BBLS)	Exploration well no.	Field size (oil 10^6 BBLS)
3	28	56	177	141	8.8
7	26	57	43	144	29
8	775	58	33	145	450
9	114	62	178	152	5.9
11	31	71	15	161	8.8
12	337	76	22	162	49
17	41	78	11	171	100
18	113	81	8.1	174	10
20	1328	82	35	176	8.8
21	21	88	25	178	17
22	13	90	170	—	—
29	455	92	19	189	12
33	89	102	56	195	125
34	482	105	42	197	20
35	70	109	335	202	8.8
39	215	110	21	203	8.8
45	62	116	50	209	6.9
46	58	119	181	210	25
52	6.9	131	93	215	100
55	154	139	75	—	—

A plot of reserves so far discovered against number of exploration wells drilled is shown in Fig. 1.

The plot shows a declining rate of growth of known reserves with respect to amount of exploration. This phenomenon, which has long been recognized in the

* Petroleum province – a geographic area known to have commercial hydrocarbon accumulations.

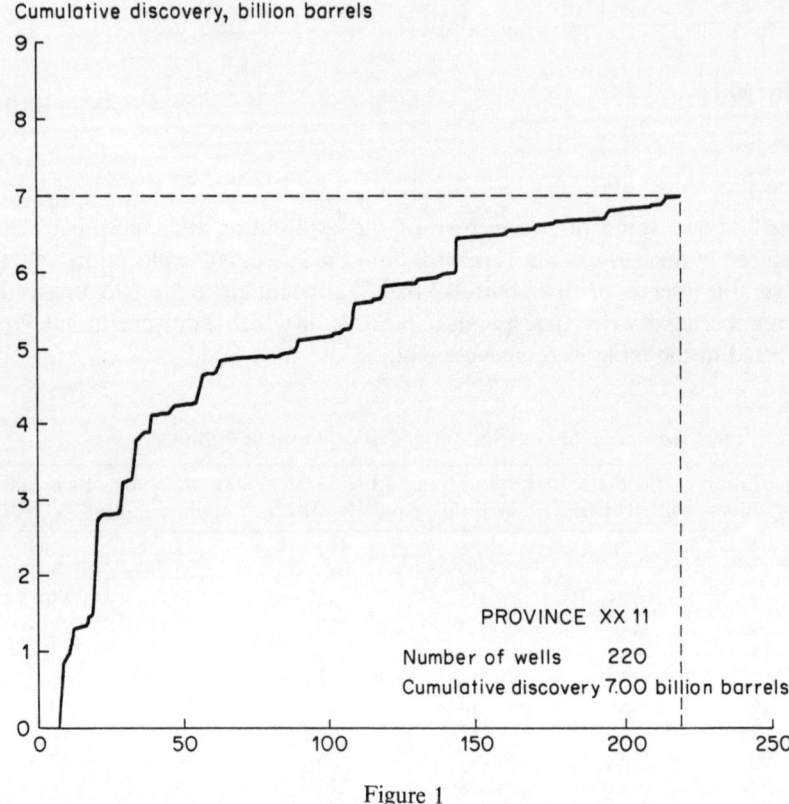

Figure 1

oil industry, is thought to be attributable to two factors (not necessarily equally important):

(i) A declining success rate for exploratory drillings – to be expected since there is only a finite number of fields in a province (although this effect is partly offset by a steady improvement in search techniques).

(ii) A decline in the sizes of the fields that are discovered – to be expected because oil companies are able to find, through their technological and geological expertise, the largest fields early in exploration and, as a result, come to progressively smaller ones in later stages.

Analyse the data with a view to finding and evaluating a statistical model to describe the process.

[Table 1 and Figure 1 are reproduced by kind permission of The Editors, *Journal of the Royal Statistical Society*, Series A.]

*   *   *

Level: 3   *Duration:* 2 weeks or more (see below)   *Type:* Individual

Comment: The data come from Meisner and Demirmen (1981), where some
further background information will be found and an account of a
Bayesian analysis of the data. There is no particular benefit in
exactly following Meisner and Demirmen's analysis, and so the
teacher might well not wish to draw attention to the paper initially,
or, if he does draw attention to it, he might suggest that an
alternative analysis be found. Two broad approaches are possible.
One is to look for a purely descriptive model without considering
underlying causes. Some form of curve-fitting would be a natural
technique to apply in this case. The alternative is to try to construct a
model which takes account of mechanisms. The two factors above
are thought to be important: further background reading may, of
course, suggest others to the student. This second, explanatory, kind
of model is likely to be most useful in practice, since extrapolation
with it will be on a firmer basis. For this reason, and because it will
engage more of the student's practical abilities, the teacher might
wish to steer his class towards it. A thorough investigation in this
direction will almost inevitably be iterative, as models are checked
and refined, and so may require more than the two weeks that would
be enough for empirical modelling.

APPENDIX

The following is the text of a document given to statistics students at the University of Sheffield.

The writing of reports

A.1 INTRODUCTION

For most, possibly all, of the practical projects that you will carry out in your course you will be required to write a report, summarizing what you have done and what your conclusions are: the suggestions in this handout are intended to help you with that task. (This material is concerned particularly with the relatively short report appropriate to small projects, but most of the points apply much more widely – to theses for postgraduate degrees, for example, and to reports to be written within the context of ordinary commercial employment. There may well be a required or preferred form for these, which you should then study and follow, but this is likely to dictate only the structure, and the other aspects will then be similar to the present ones.)

There are several aspects, quite different in kind, of writing a report, and they will be discussed in turn:

(a) structure;
(b) content;
(c) presentation;
(d) style;
(e) the preparation in detail.

Only general guidelines can be given, of course, which must then be interpreted and applied to the immediate task. The most important thing by far is to realize that a report is good if – and only if – it is effective, which means that it gets its message across: anything which diminishes its clarity or tends to irritate the reader will detract from that.

A.2 STRUCTURE

There are several possibilities, but the one suggested here has much to commend it; the general approach is similar to that of Ehrenberg (1982). The report will be made up of several components, listed below in the order *in which they should appear*. Some of them may well be better split up into sections, while some reports,

184

perhaps because of their brevity, will not require as full a structure as that suggested here. Something that should certainly be avoided is the narrative approach, of the kind that says 'first I did this and found that; next I did something else . . .', because this would make the report difficult to follow.

(a) Title

Every report needs a title, preferably on a separate title page, which should also include the author's address, department or other affiliation, and office (chairman, secretary, etc.) if he has one and if it is relevant, and the date.

(b) Summary

A brief summary – the shorter the better, and certainly no more than half a page – should follow, to tell the reader what the report is about, in general terms, and why it may be worth reading. For example:

> Some data relating Sheffield house prices to various geographical and physical characteristics, previously analysed by linear regression, are here analysed by generalized linear model (GLM) methods, and the results compared with those found earlier.

The summary should not, other than in exceptional circumstances, contain symbols, or numbers, and should contain technical terms only if they are widely understood.

(c) Introduction

This should describe the general context and background, including a (general) description of what data are available, and the aims of the investigation, together with some indication of the methods used. Overlap with the summary should not be a worry: the summary may in some circumstances be reproduced and used separately. It is difficult to generalize about length, but in the present context anything longer than about two sides would be surprising.

(d) Results, conclusions, and recommendations

Here it should be the *main* results and conclusions only that are given. Depending to some extent on their nature, subsidiary results and deductions may be gathered into a separate component, be combined with the description of methods, or be put into an appendix.

(e) Methods

These, which include any theory that is necessary, should now be described in reasonable detail. How much detail is always difficult to decide, but if one tries to write it as though for someone of comparable standing – a student in the same year, for example – who does not know anything about the subject of the report in detail, in such a way that he could repeat the study, this will be a good guide. Here again, do not be afraid of putting very detailed and long descriptions into one or more appendices.

What is quite likely is that this component will need to be split into a number of sections: a preliminary analysis, using graphical and simple descriptive methods, and then a full-scale analysis, for example. Note that what is being suggested here is that the techniques you need should be described in this section, whereas the results of using them will have been described in the previous one. This may at first sight seem an unnatural procedure, but you will find that it makes for efficient communication of what you found and how you found it, and that, after all, is a main aim of the report.

(f) General discussion

It may be appropriate to give some account of previous investigations of the same or related problems, or to relate the present conclusions to others in a connected area. It very likely *is* appropriate to discuss how far the original aim was successfully achieved, and if it was not, why not; to say what you would do differently if you had to begin again; and what more might be done to carry things further. Reservations about the quality of the data, and so on, also belong here.

(g) References

The usual style in learned journals and books in statistics and mathematics is to refer to sources in the text by name and date – as was done above at the beginning of this main section – and then to collect them in an alphabetical list at the end: see an example at the end of this document. In other contexts a different convention may apply, and you should follow whatever is the custom.

(h) Appendix

Whatever does not fit naturally earlier should go into an appendix. In particular, computer program listings, which it will be appropriate to give if programs are individually written (and at all complicated), should be relegated to an appendix, possibly accompanied by a structure plan or flow chart, and *annotated*, so that the reader can follow them. Small, and small numbers of, tables can appear in the main report, but otherwise an appendix is likely to be the right place.

A.3 CONTENT

The decisions about the detailed contents are of course extremely important, but it is difficult to give much in the way of general guidelines; nevertheless a few things which are universally applicable may be said.

(a) A report must be self-contained – everything should either be stated explicitly, be justified by a reference to some other source, or be common ground (obvious, or common sense, or just well known for the intended readership). In particular, the source of any data should be clearly indicated.
(b) It is not usually sensible to describe everything that you did in detail. If you decide that one of your approaches to a question is wrong or inappropriate, then, unless you are convinced that most others would begin by supposing it acceptable, it should be omitted completely. If your initial approach is clearly included in a later approach, no more than (at most) a passing reference to the former is necessary. For instance:

> Regression on temperature showed some interesting features, but it became clear that both temperature and pressure should be brought into the model [and the results are as follows].

A.4 PRESENTATION

Reports should be attractive, and easy to find one's way through. It follows that they should be neat, clean, and tidy, and that they must be legible – so that if they are handwritten you should try to ensure that your handwriting is not a problem.

Pages, tables, diagrams, etc., should be numbered, as should sections, which should have headings; headings should stand out (by being underlined, or on a separate line, for example). In fact one can think of a report as needing signposts – partly to make it clear where one is (section headings, etc.), and partly to show where one is going (which will be done either by the wording or by dividing the report into subsections, each with a heading, or both). For example:

> Thus, in the light of the analysis, it seemed worth investigating whether the variance of X was constant, and also attempting to model the relation between Y and Z.
>
> 6.1.1 The variance of X.
>   . . .

In much the same way, graphs need to be neatly drawn, carefully labelled and titled; scales should be chosen *sensibly* and the units indicated. A freehand graph, carefully drawn, may be appropriate; a sketch graph with various amendments, and axes carelessly drawn, will never be.

Tables, again, need titles, need to be boxed in, and figures in columns should be aligned correctly (usually so that the decimal points fall in the same line.) Rows and columns usually need labels (headings), and units must be stated in, or perhaps close to, the table.

Computer output needs some thought. For a student project report it may not be necessary to copy the results from the output so that they conform in appearance with the rest of the report (though it probably would be, at least for line printer output, for more serious purposes); but they will certainly need editing, to remove superfluous results, control commands, and so on, and it will probably be necessary to improve the labelling.

You may well find Chapman and Mahon (1986) helpful in these matters.

A.5 STYLE

(a) Earlier exhortations to be brief can stand repetition here. Closely connected with this is the advantage of using shorter rather than longer words when possible, and of avoiding jargon: if 'educational establishments' means only 'schools and colleges', then the latter is much to be preferred.

(b) Use of the personal style ('I'), needs to be carefully done. (It is virtually never appropriate to use 'you'). It is in principle reasonable – since it describes the case exactly – when judgement is involved: 'I consider that the plot is sufficiently nearly a straight line that the data may be taken as normal.' It sounds rather naive if used in a narrative context: 'I carried out a regression analysis.' In either case a reader may be irritated or distracted by it, since it is unusual, which will then detract from the impact of the report.

The indefinite pronoun 'one' can be used, but becomes tiresome very quickly, and the most usual solution is to make great use of impersonal verbs and of the passive voice. 'A regression analysis was carried out.' 'It can be seen from the graph . . .'.

In mathematical-style arguments it is customary to use 'we': 'We can now see . . .'. But the implication is that the reader and the author are doing this together, so that 'We carried out a regression analysis' is incorrect (unless there are two or more authors, to whom 'we' then refers).

(c) Some people would not agree, but there are many who care about spelling, and, in consequence, incorrect spellings may prove a distraction to your reader and a hindrance to efficient communication. If you know that your spelling is shaky, therefore, be prepared to look words up in the dictionary, and to remember corrections for the future. Many people cannot spell 'separate', and words like 'recommend', with some but not all letters doubled, need care.

(d) Be careful not to use words inappropriately: for example the similar sounding words 'trend' and 'tendency' are not interchangeable. Do not use the word 'significant' in a non-technical way in the middle of a statistical report.

(e) Grammar and punctuation are, again, areas which give trouble. Paragraphs divide the text into manageable pieces, usually with some unifying theme. One stage down from this is the sentence, ending with a full stop (or question mark). Sentences have to be complete in themselves, so that the following are not acceptable:

'Which prove that it has no influence.'
'A standard analysis is possible. By regression.'

(f) Style in a general sense is concerned as much with how the text flows – how it sounds – as with anything else, and if you read your report to yourself, pausing firmly at full stops, less so at commas, and so on, you may be able to detect grammatical errors connected with punctuation. You will also discover that using the same word time and time again, or starting two or three sentences in succession in the same way gives a monotonous effect; and you will also find that a succession of very short sentences breaks up the flow in an unpleasant manner.

Style is, in the end, something that you can develop only by practice, and by critically rereading what you have written. The famous maxim quoted by Dr Johnson:

> Read over your composition, and where ever you meet with a passage which you think is particularly fine, strike it out.

illustrates an appropriately self-critical attitude of mind, even if it exaggerates the action usually needed.

You may find *The Complete Plain Words* (Gowers, 1973) and *A Dictionary of Modern English Usage* (Fowler, 1983) (more idiosyncratic but more fun) useful as references on grammar, punctuation, and style.

A.6 THE PREPARATION IN DETAIL

There are really only two things to be said here: it is essential to plan the report before you start writing, and it is also essential to (be prepared to) revise it, possibly extensively, after you have written the first version. This handout, for example, was certainly revised noticeably.

Exactly what is worth doing depends on the scale, both of time and of the report – whether it is to be 5 pages or 50. If there is time, however, it is worth leaving a period between writing the first attempt and rereading it; if you can persuade someone else to read it and give you his reactions, so much the better, since he is going to be in a similar position to the final reader, whereas you will know what you meant to say, even if you did not actually say it.

If the report is to describe an activity which is extended in time, it is usually sensible to write the first draft piecemeal, as the information becomes available, to reduce what may be a formidable task if left entirely to the end. Also, you may find that the act of writing itself suggests new ideas and so may influence your plans for later parts of the analysis.

Note, finally, the rather obvious point that the various sections of a report need not be written in the order in which they will finally appear: it will often be easier to write the Introduction and Conclusions or Recommendations *after* making decisions about the contents of later sections and writing them.

REFERENCES

EHRENBERG, A. S. C. (1982) Writing technical papers or reports. *The American Statistician*, **36**, 326–329.

CHAPMAN, M. in collaboration with Mahon, B. (1986) *Plain Figures*. HMSO, London.

FOWLER, H. W. (1983) *A Dictionary of Modern English Usage* (2nd edition revised by Sir Ernest Gowers). Oxford University Press, Oxford and New York.

GOWERS, SIR ERNEST (1973) *The Complete Plain Words* (2nd edition revised by Sir Bruce Fraser). Penguin Books, Harmondsworth, Middlesex.

References

ALLISON, J., and BENSON, F. A. (1983) Undergraduate projects and their assessment. *Proc. IEE*, **130**, Part A, No. 8, 402–419.

AMERICAN STATISTICAL ASSOCIATION COMMITTEE ON TRAINING OF STATISTICIANS FOR INDUSTRY (1980) Preparing Statisticians for careers in industry (with discussion). *The American Statistician*, **34**, 65–75.

ANDREWS, D. M., and HERZBERG, A. (1985) *Data: a Collection of Problems from Many Fields for the Student and Research Worker*. Springer-Verlag, New York.

ARMITAGE, P. (1977) Further thoughts on curricula. *Roy. Statist. Soc. News and Notes*, June, 6–7.

BACON, F. (1625) *The Essays, or Counsels, Civil and Moral* of Francis Bacon, Lord Verulam. (Everyman edition, Dent, London, 1906.)

BARON, J. S., and WORSDALE, G. J. (1980) *Assignments in Quantitative Methods*. Polytech Publications Ltd., Cheshire.

BISSELL, A. F. (1975) The use of case-studies in teaching statistics. *Bulletin in Applied Statistics*, **2**, 29–32.

BOEN, J. R. (1982) Discussion of Zahn (1982a). pp. 537–540 in Rustagi and Wolfe (1982).

BOEN, J. R., and ZAHN, D. A. (1982) *The Human Side of Statistical Consulting*. Lifetime Learning Publications, Belmont, California.

BONINGTON, C. (1976) *Everest The Hard Way*. Hodder and Stoughton, London.

BOX, G. E. P. (1984) The importance of practice in the development of statistics. *Technometrics*, **26**, 1–8.

BRADLEY, R. A. (1982) The future of statistics as a discipline. *J. Amer. Statist. Assoc.*, **77**, 1–10.

BROWN, P. J., and PAYNE, C. D. (1986) Aggregate data, ecological regression, and voting transitions. *J. Amer. Statist. Assoc.*, **81**, 452–460.

CHAPMAN, D. G. (1978) The plight of the whales. pp. 105–111 in Tanur *et al.* (1978).

CHAPMAN, M. in collaboration with Mahon, B. (1986) *Plain Figures*. HMSO, London.

CHATFIELD, C. (1982) Teaching a course in applied statistics. *Appl. Statist.*, **31**, 272–289.

CLEMENTS, L., and CLEMENTS, R. R. (1978) The objectives and creation of a course of simulation/case studies for the teaching of engineering mathematics. *Int. J. Math. Educ. Sci. Technol.*, **9**, 97–117.

COX, D. R., and SNELL, J. (1981) *Applied Statistics: Principles and Examples*. Chapman & Hall, London.

CREWE, I., and FOX, A. (1984) *British Parliamentary Constituencies and Statistical Compendium*. Faber & Faber, London.

CROASDALE, R. (ed.) (1985a) *Student Projects in Statistics*. (Proceedings of an ASLIP Conference.) ASLIP (Association of Statistics Lecturers in Polytechnics), England.

CROASDALE, R. (1985b) The assessment of projects within a mathematics degree. pp. 14–20 in Croasdale (1985a).

CROSS, M., and MOSCARDINI, A. O. (1985) *Learning the Art of Mathematical Modelling*. Ellis Horwood, Chichester.

DES (Department of Education and Science) (1985) An assessment of the costs and benefits of sandwich education. (Report of a Committee on Sandwich Education). HMSO, London.

191

DUNCAN, K. D. (1966) Effects of an artificial acclimatization technique on infantry performance in a hot climate. *Ergonomics*, **9**, 229–244.

DYER, K. (1984) Catching up the men. *New Scientist*, 2nd August, 25–26.

EFRON, B. (1982) Maximum likelihood and decision theory (1981 Wald Memorial Lecture). *Ann. Statist*, **10**, 340–356.

EHRENBERG, A. S. C. (1982) Writing technical papers or reports. *The American Statistician*, **36**, 326–329.

FELLER, W. (1968) *An Introduction to Probability Theory and its Applications*, Vol. 1 (3rd edition), J. Wiley, New York and London.

FINNEY, D. J. (1974) Problems, data and inference. *J. Roy. Statist. Soc.* A, **137**, 1–22.

FINNEY, D. J. (1982) The questioning statistician. *Statistics in Medicine*, **1**, 5–13.

FIRTH, D. (1982) Estimation of voter transition matrices from election data. London University MSc Thesis.

FISHER, R. A. (1953) The expansion of statistics. *J. Roy. Statist. Soc.* A, **116**, 1–6.

FOWLER, H. W. (1983) *A Dictionary of Modern English Usage* (2nd edition revised by Sir Ernest Gowers). Oxford University Press, Oxford and New York.

FREEMAN, D. H. JR., GONZALEZ, M. C., HOAGLIN, D. C., and KILSS, B. A. (1983) Presenting statistical papers. *The American Statistician*, **37**, 106–110.

GORDON, A. (1974) Accuracy of weather forecasts. *Nature*, **252**, 294–295.

GOWERS, SIR ERNEST (1973) *The Complete Plain Words* (2nd edition revised by Sir Bruce Fraser). Penguin Books, Harmondsworth, Middlesex.

GREEN, F. H. W. (1975) The February–June weather relationship in north-west Europe. *Nature*, **253**, 522–523.

GREENSTONE, M. H. (1979) Spider feeding behaviour optimizes dietary essential amino acid composition. *Nature*, **282**, 501–503.

GREENSTONE, M. H. (1981) Rejoinder. *Nature*, **290**, 165–166.

GREENWOOD, J. J. D. (1981) Does spider feeding behaviour optimize dietary essential amino acid composition? Letter to Editor, *Nature*, **290**, 165.

GRIFFITHS, J. D., and EVANS, B. E. (1976) Project work in statistics. *The Statistician*, **25**, 117–123.

HAWKES, A. G. (1969) An approach to the analysis of electoral swing. *J. Roy. Statist. Soc.* A, **132**, 68–79.

HAWKES, A. G. (1980) Teaching and examining applied statistics. *The Statistician*, **29**, 81–89.

HEALY, M. J. R. (1973) The varieties of statistician. *J. Roy. Statist. Soc.* A, **136**, 71–74.

HILL, I. D. (1974) Association football and statistical inference. *Applied Statistics*, **23**, 203–208.

HOWARD, K., and SHARP, J. A. (1983) *The Management of a Student Research Project*. Gower Press, Aldershot.

HUNTER, W. G. (1981) The practice of statistics: the real world is an idea whose time has come. *The American Statistician*, **35**, 72–76.

ISI (1986) International Statistical Institute declaration on professional ethics. *International Statistical Review*, **54**, 227–242.

JONES, J. D., and KANJI, G. K. (1980) The role of professional experience in statistical education. *The Statistician*, **29**, 196–203.

JOWETT, G. H., and DAVIES, H. M. (1960) Practical experimentation as a teaching method in statistics (with discussion). *J. Roy. Statist. Soc.* A, **123**, 103–105.

KANESTRØM, I. (1975) February–June weather relationships in Norway. *Nature*, **257**, 337.

KANJI, G. K. (1979) The role of projects in statistical education. *The Statistician*, **28**, 19–27.

KANJI, G. K. (1983) Practicals, case-studies, projects and professional experience: some methods of developing skills among budding statisticians. *Proc. First International Conference on Teaching Statistics*, 671–684.

KENDALL, M. G. (1968) On the future of statistics – a second look (with discussion). *J. Roy. Statist. Soc.* A, **131**, 182–204.

KEYFITZ, N. (1978) How crowded will we become? pp. 355–367 in Tanur *et al.* (1978).

KITSON, T. (1984) The ultimate mile. *New Scientist*, **103**, 34.

LAMB, H. (1950) Types and spells of weather around the year in the British Isles. *Q. Jour. Roy. Met. Soc.*, **76**, 393–429.

LI, C. C. (1964) *Introduction to Experimental Statistics*. McGraw-Hill, New York.

McCARTHY, C., and RYAN, T. M. (1977) Estimation of voter transition probabilities from the British general elections of 1974. *J. Roy. Statist. Soc.* A, **140**, 78–85.

McCULLAGH, P., and NELDER, J. A. (1982) *Generalized Linear Models*. Chapman & Hall, London.

MARDIA, K. V. (1972) *Statistics of Directional Data*. Academic Press, New York.

MEISNER, J., and DEMIRMEN, F. (1981) The creaming method: a Bayesian procedure to forecast future oil and gas discoveries in mature exploration provinces. *J. Roy. Statist. Soc.* A, **144**, 1–31.

MORONEY, M. J. (1951) *Facts from Figures*, Pelican, Harmondsworth.

MOSTELLER, F. (1980) Classroom and platform performance. *The American Statistician*, **34**, 11–17.

MULLER, M. E. (1978) Information, simulation and production. pp. 458–465 in J. M. Tanur *et al.* (1978).

PEARSON, E. S. (1971) The Department of Statistics, 1971: a year of anniversaries. Lecture given at University College, London, December 1971.

PEARSON, E. S. (1978) Statistics and probability applied to problems of anti-aircraft fire in World War II. pp. 474–482 in Tanur *et al.* (1978).

PEARSON, K. (1978) *A History of Statistics in the 17th and 18th centuries* (ed. E. S. Pearson). Griffin, London.

PHILLIPS, D. P. (1978) Deathday and birthday: an unexpected connection. pp. 71–85 in Tanur *et al.* (1978).

PRIDMORE, W. A. (1985) The market. pp. 2–11 in *Whither Statistical Education*. Centre for Statistical Education, Sheffield.

REEP, C., and BENJAMIN, B. (1968) Skill and chance in association football. *J. Roy. Statist. Soc.* A, **131**, 581–585.

REEP, C., POLLARD, R., and BENJAMIN, B. (1971) Skill and chance in ball games. *J. Roy. Statist. Soc.* A, **134**, 623–629.

RENDALL, F. J., and WOLF, D. M. (1983) *Statistical Sources and Techniques*. McGraw-Hill (UK) Ltd., Maidenhead.

ROYAL STATISTICAL SOCIETY (1986) Report of a joint working party of the Royal Statistical Society and the Institute of Statisticians on statisticians: supply and demand. *J. Roy. Statist. Soc.* A, **149**, 122–145.

RUSTAGI, J. S., and WOLFE, D. A. (eds.) (1982) *Teaching of Statistical Consulting*. Academic Press, New York and London.

SCHOFIELD, A. N. (1959) *Parliamentary Elections* (3rd ed.). Shaw, London.

SCOTT, J. F. (1976) Practical projects in the teaching of statistics at universities. *The Statistician*, **25**, 95–108.

SEARLE, S. R. (1971) *Linear Models*. J. Wiley, New York and London.

SIMS, G. D. (1976) Electrical engineers – education and the future. *J. Royal Signals Inst.*, **12**, 149–172.

TANUR, J. M., MOSTELLER, F., KRUSKAL, W. H., LINK, R. F., PIETERS, R. S., RISING, G. R. and LEHMANN, E. L. (eds.) (1972) *Statistics: A Guide to the Unknown*. 2nd edn. Holden-Day, San Francisco.

THOMPSON, M. (1975) On any given Sunday: fair competitor orderings with maximum likelihood methods. *Jour. Amer. Statist. Assoc.*, **70**, 536–541.

TUKEY, J. W. (1963) Contribution to discussion following paper by G. W. Bennett and E. A. Cornish: A comparison of the simultaneous fiducial distributions derived from the multivariate normal distribution. *Bull. Inst. Int. Statist.*, **40**, 902–939.

WATMAN, M. (1981) *Encyclopaedia of Track and Field Athletics*. St. Martin's Press, New York.

WATTS, D. G. (1981) A task-analysis approach to designing a regression analysis course. *The American Statistician*, **35**, 77–84.

WHITNEY, C. A. (1978) Statistics, the Sun and Stars. pp. 450–457 in J. M. Tanur *et al.* (1978).

WOODWARD, WAYNE A. and SCHUCANY, WILLIAM R. (1977) Bibliography for statistical consulting. *Biometrics*, **33**, 564–565.

YATES, F. (1968) Theory and practice in statistics. *J. Roy. Statist. Soc.* A, **131**, 463–477.

ZAHN, DOUGLAS A. (1982a) Teaching statistical consulting: statistical and non-statistical aspects. pp. 517–536 in Rustagi and Wolfe (1982).

ZAHN, DOUGLAS A. (1982b) The evolution of supervised statistical consulting at Florida State University: A Response to the Review of James R. Boen. pp. 541–548 in Rustagi and Wolfe (1982).

Author Index

Subject Index